THE SPIRITUAL SIGNATURE OF OUR TIME

IN THE ERA OF CORONAVIRUS

The School of Spiritual Science

Edited by
Ueli Hurter and Justus Wittich

RUDOLF STEINER PRESS

Translated by Christine Howard

Rudolf Steiner Press
Hillside House, The Square
Forest Row, East Sussex RH18 5ES

www.rudolfsteinerpress.com

Published by Rudolf Steiner Press in 2021

Originally published in German under the title *Coronazeit: Zur geistigen Signatur der Gegenwart* by Verlag am Goetheanum Dornach, Switzerland, in 2021

© Verlag am Goetheanum 2021
This translation © Rudolf Steiner Press 2021

This book is copyright under the Berne Convention. All rights reserved. Apart from any fair dealing for the purpose of private study, research, criticism or review, no part of this publication may be reproduced, stored in a retrieval system, or transmitted in any form or by any means, electronic, electrical, chemical, mechanical, optical, photocopying, recording or otherwise, without the prior written permission of the copyright owner. Inquiries should be addressed to the Publishers

The rights of the individual contributors to be identified as the authors of this work have been asserted in accordance with sections 77 and 78 of the Copyright Designs and Patents Act, 1988

A catalogue record for this book is available from the British Library

ISBN 978 1 85584 594 7

Cover by Morgan Creative featuring artwork by Wolfram Schildt
Typeset by Symbiosys Technologies, Visakhapatnam, India
Printed and bound by 4Edge Ltd., Essex

Contents

Foreword — 1
Ueli Hurter and Justus Wittich

Ecology and the Pandemic
What can we learn from COVID-19? — 3
Georg Soldner

We Poor Royal Children
Corona and the Social Challenges of Our Time. — 24
Gerald Häfner

How Is Our Behaviour Mirrored In The Ecosystem?
Prevailing Perspectives in Agriculture. — 40
Jean-Michel Florin

'Building a Bridge to Right-Wing Extremism?'
On Anthroposophy in the Time of National Socialism — 57
Peter Selg

Can what is foreign be understood? — 103
Constanza Kaliks / Paula Edelstein

Are we making a religion out of science? — 111
Matthias Rang

Goetheanistic Aspects in Dealing with COVID-19 — 121
Johannes Wirz

Individual Responsibility in Corona Times — 133
Stefan Hasler / Ueli Hurter

Digital Challenges in Education — 141
Florian Osswald / Claus-Peter Röh

The Hidden Sun 159
Christiane Haid

What effects do inner work and meditation have
on the healing powers of the human being? 179
Matthias Girke

What are the Intentions of the School of Spiritual
Science at the Goetheanum? 198

About the Authors 201

Foreword

Although the Goetheanum as the School of Spiritual Science presented a first differentiated compilation, *Perspectives and Initiatives in the Times of Coronavirus*, very early on in May 2020 (German original) from the perspective of its various specialized sections, in the following months anthroposophists were often brought into widespread alignment with 'corona deniers', conspiracy theorists and opponents of vaccination in the media. Although this happened here and there, anthroposophy is about a path of knowledge that goes beyond natural science to include soul and spirit. There are irrefutably many contemporary phenomena that need to be critically reflected on, and above all, other approaches towards solutions need to be recommended or reference made to existing tried-and-tested practice in this regard. This is enlightenment in the best possible sense of the word!

In response, ten lectures were held at the Goetheanum and this book contains the transcriptions of these. These evenings were conceived as a series and held under the title *The Signature of Our Time*. The audience was either a local or worldwide one, as the lectures were recorded on video and made available online after a few days, in German and with English voice-overs.

These lectures took place between 12 October and 21 December 2020, which represented a particular period in an exceptional year. It happened during the time of the emergence of the second Corona wave. On the first evening, there was still an almost normal world accompanied by relaxations of measures and questions of whether there would even be a second wave. All that was required was an appropriate protection/safety concept for holding a public event. By the second evening, the situation had already changed dramatically; the numbers of positive tested cases had increased exponentially. A short time later, Solothurn, the canton where the Goetheanum is based, then allowed only a maximum of 30 people to attend an event. Finally, by December, all public and cultural gatherings had been banned, so that the last three lectures took place without an audience, only 'face to face' via the camera. Thus the Goetheanum, with this lecture series, now entered the digital world. A stroke of good luck or of bad luck?

When the series was conceived in early September 2020, the first Corona wave was considered a thing of the past. Society was increasingly in uproar, mirrored, for instance, in large demonstrations. How real is this pandemic? Is the COVID-19 disease worse than flu or not? Does the state have the authority to suspend fundamental rights for prolonged periods, or is it important to rebel against it out of civil courage? Is science omniscient and omnipotent—or is it possible to judge the situation out of our own sense of responsibility? 'The pandemic is changing our lives quite radically' was the basic attitude to life during these weeks. This was also unveiled through the fact that cracks in the fabric of society were appearing where none had existed before—front lines and images of the 'enemy' were being formed: in families, in institutions such as schools, or in public opinion and in most countries, sharp contrasts had emerged.

In this situation the Goetheanum did not want to remain silent. The series of lectures by the Section Leaders was intended to bring forward contributions that could help towards more discerning views for making judgements, which could lead to greater individually-based responsible capacities for action. They are about thought-provoking impulses, about larger contexts and about an attempt to take into consideration the various dimensions of civilization that have been shown to be problematic through the Corona time.

Finally, in February 2021, we decided to transcribe the lectures—updated with current facts—and publish them as a book, again in German and English. The reason for this is not least that the Goetheanum, as a School of Spiritual Science and a public cultural centre, has had to officially close again from mid-December 2020 to the end of March 2021. So we are making a virtue out of necessity by holding digital events, publishing this book and the weekly journal *Das Goetheanum* (German only). This fortunately has brought us into contact with a whole new audience, people from all regions of the world who would never have been able to travel to Switzerland and the Goetheanum, even in normal times. In this sense we are offering this book, as a contribution to the dialogue in this difficult, but at the same time awakening contemporary situation.

Ueli Hurter and Justus Wittich
Spokespersons for the Goetheanum Leadership
Dornach, 8 March 2021

Georg Soldner

Ecology and the Pandemic: What can we learn from COVID-19?

Different Backgrounds of Experience

What can we learn in our current joint experience of the COVID-19 pandemic? Conversations about how to deal with this novel infectious disease, COVID-19, quickly threaten to become emotional. I recently spoke with a doctor friend and school doctor who herself had contracted COVID-19. She did not have to go to hospital, she was treated with great care, but her illness and recovery took not one, not two weeks, but several months, as is the case with many who, statistically, are not considered severe cases because they can be treated at home. Her lungs as well as her cardiovascular system had been attacked and weakened—she reports a deeply profound experience of weakness—and only very slowly could she resume everyday activities. She experienced that the course of this disease is perhaps also coloured by how exhausted one's own life forces were before becoming infected, which occurred when an infected person nearby coughed on her. On the other hand, she clearly experienced the effectiveness of external compresses and nature-based, anthroposophic medicines, whereas for her there were hardly any meaningful effective treatments on offer by conventional medicine. She reported, 'When I now talk about the course of my illness, the joint discussions with parents and teachers at school are no longer problematic.'

An exchange about COVID-19 is initially not always easy and depends on the respective location and extent of experience. I recently spoke with heads of large special-needs educational institutions in Germany and the Netherlands. Our Dutch colleague reported, among other things, about a 39-year-old special-needs educational therapist who died of COVID-19 in September. Our German colleague spoke about a facility with 140 people with assistance needs, some of whom also require intensive physical care, and about the same number of staff who were then completely quarantined with several cases of illness. Especially where young staff members have to be there for people with disabilities or for older people, it is not so easy to implement the often-mentioned protection of risk groups, over the long term.

My own testimony of this pandemic continues to be shaped by an early contact with the family of a Milanese colleague and friend who, like more than 150 other doctors in upper Italy, died of COVID-19 in the early days of the pandemic. I emphasize the number of Italian colleagues who died, because we haven't experienced this in any influenza epidemic. My colleague was treated in intensive care for four weeks with great dedication, and finally died of a secondary infection, without having any contact with his wife or children during this time. In Catholic Italy, and not only there, both the living and the dying are isolated to an extent that seemed unimaginable not so long ago. He experienced his last weeks of life and his death without a single visitor, without a single farewell. A funeral service was not possible in Milan either. A few weeks later, his urn was delivered to his family.

If we look at South America and learn about it from images of places like Manaus or Ecuador, we can observe that in many places the conditions of medical care for the population were and are difficult to describe. The chances of treatment are very unequally distributed there, and life expectancy in a Brazilian favela, even without COVID-19, is already 18 years less than the life expectancy of those born in one of those Brazilian residential areas where every house is surrounded by a wall and access is guarded around the clock. Even before COVID-19, there was a strange—one could even call it inhumane—coexistence of maximum control and absence of control, as the poor in these countries are exposed to unparalleled levels of violence and organized crime. Access to the health system is very limited for the poor of South America, except for their vaccination programmes. Lack of healthy food and clean water, while breathing in polluted air, is the reality for the poor. Air pollution has been shown worldwide to be a risk factor for severe courses of COVID-19.

COVID-19 is a disease that particularly attacks the respiratory and cardiovascular systems, the middle sphere of the human organism. However, it also attacks the middle sphere of society threatening to split us in our dialogue with one another. So when we speak here, from Switzerland, about a global pandemic, its origin, its character and what we can learn from it, it is important to note that we are dealing with a reality that first of all begs our utmost respect and our compassion towards all those affected. Perhaps this is the first thing we can learn from this pandemic: that this pandemic and its consequences can affect people in very different ways, and that this has to do not only with the virus, but with us, with how we engage with each other socially and economically, with what educational opportunities are offered to each person, and with what attitude all living beings that live together on this planet are treated.

How could this pandemic have happened?

Significantly, an early, very accurate prediction of the Corona pandemic can be found in the German Bundestag's 'Risk Analysis for Population Protection', January 2013.[1] This deals with ecological issues, firstly with the issue of extreme meltwater and then from page 55 onwards, with a pandemic called the 'Modi-SARS virus':

> The present scenario describes an extraordinary epidemic event based on the spread of a novel pathogen. The scenario is based on the hypothetical pathogen 'Modi-SARS', (...) which is very closely related to the SARS [Corona] virus. (...) The incubation period—i.e. the time from the transmission of the virus to a person until the first symptoms of the disease appear—is usually three to five days, but can range from 2 to 14 days. (...) The symptoms are fever and dry cough, the majority of patients have shortness of breath, changes in the lungs visible in X-rays, chills, nausea and muscle pain. (...) The severity varies in different age groups. Children and adolescents usually have milder courses of disease with fatality rates of around 1%, while the fatality rate for those over 65 years of age is 50%.

In fact, according to a recent study, it is 22.3% in over 85-year-olds and almost 0% in children.[2] Precise assumptions about the duration of the disease follow. Further:

> The event begins in February, in winter, in Asia, but its significance is only recognized there a few weeks later. (...) In April, the first identified 'Modi-SARS' case occurs in Germany.

With a difference of two months, this coincides exactly with the 2020 pandemic.

> The pathogen originates in Southeast Asia, where the pathogen found in wild animals is transmitted to humans via markets. Since the animals themselves do not become ill, the risk of infection is not immediately apparent.[3]

Yes, SARS-CoV-2 originates from wild animals. They are kept in cramped cages, sold at 'wet markets' and often slaughtered on site while many people are present in crowded confined spaces, including those working there, all of whom come into direct contact with the animals that are stressed to the maximum. The report thus identifies conditions that today appear relatively certain as necessary preconditions for this pandemic, which also pose a source of real danger in the future, possibly from even much more dangerous ones, through wild animals being kept in cramped confinement, their lives acutely threatened, their vitality often severely impaired, coupled with the unnatural proximity of various species, for instance, bats and pangolins, as well as humans and other animals. In this situation, the immune systems of the animals themselves are severely impaired, viruses can vigorously multiply in the animals, double infections are possible and, as a result, the formation of a so-called chimera from two different viruses, which can, for example, facilitate the transition of a virus from one animal species to another species and to humans. SARS-CoV-2 may well have been created in this way. Finally, the transition from animal-specific viruses to humans and back again, as has already been demonstrated, for example, in minks in Dutch and Danish fur farms last spring, ultimately leading to the culling of 17 million mink in Denmark after a new SARS-CoV2 virus mutation was detected there. Biologists cynically refer to these sites of extreme animal suffering as 'evolutionary accelerators'. For example, around 100 million wild animals worldwide are kept in cramped cages every year just for fur accessories and are culled after a life to which wild animals can never adapt.

The Role of Viruses
At this point, a brief word on the question of how viruses, plants, animals and humans are connected. Viruses were first discovered in plants. They accompany all living organisms. By their nature, they correspond to elements of the genome; our DNA consists to a large extent of former viral substance. Viruses facilitate changes in the genome and are an important prerequisite for evolution.[4] They can become a threat precisely when they first enter a living organism and redirect its life processes for their own reproduction. The measles virus, for example, originates from animals (cattle) and became a human pathogen. For most viruses, the threat also depends on age and thereby the immunological reactivity of the host. If a population does not know a virus at all, as was once the case with the South American populations with respect to the measles virus and perhaps the inhabitants of Amazonia with respect to the SARS-CoV-2 virus, the infection can be devastating on a wide scale.

However, the SARS-CoV-2 virus is not entirely new for many people; there is very likely a partial so-called background or adaptive immunity due to previous encounters with other corona viruses for example, which is distributed differently in the population. Because it is very difficult to assess this factor, the forecasts at the beginning of the pandemic were often overly pessimistic.

No one drew the obvious conclusions from the 2012 German Parliament's printed paper or similar documents, also presented by Bill Gates on a Ted Talk in April 2015, that humanity has to end its treatment of wild animals globally and in real terms; that fur pelts from mink and tanuki distributed on a large scale by China or Denmark, for example, and that scales and meat from pangolins, among others, must no longer be allowed to enter the market. This exhortation is irrefutable. Global wildlife protection represents an absolute priority in light of Corona. This includes, even more, the protection of soil, plants and landscapes.

This conclusion is massively reinforced when we study the circumstances under which, for example, the devastating Ebola fever broke out in Africa. The destruction of the natural habitats of wild animals drives them into unnatural proximity to humans if they are not hunted, sold and killed themselves. The Corona pandemic is part of the ecological crisis of our times, which is much broader than just a CO_2 crisis. Moreover, it is significant that meat markets and meat factories have emerged as hot spots—the excessive meat consumption of richer countries that is to say, of their people—forms the major driver of animal suffering and destruction of the environment.

Cosmic Correlations
Ecological consciousness of today focuses on the life and health of our entire planet Earth. The life of the earth, however, has its vital centre in the sun and depends on the sun's activity. This activity is stronger in summer and weaker in winter. The sun's activity is not constant, but follows an average eleven-year cycle, during which the number of sunspots fluctuates correspondingly. We know from influenza that the severity of influenza epidemics is influenced by this approximate eleven-year solar cycle. There is a virological publication from 2017 on corona viruses and sunspot activity, which shows that virus mutations are favoured by solar maximum, but also by the high-energy cosmic background radiation of solar minimum.[5] In November 2019, N.C. Wickramasinghe published a pandemic

warning explicitly based on the changes in solar activity and increased cosmic background radiation:

> In fact, new viral infections are predictable. In a letter entitled 'Space weather and pandemic warnings?' in *Current Science* published in November 2019, we explicitly reminded the world that new infectious viruses are likely to emerge in the coming months during the lowest solar minimum of the sunspot cycle in over a hundred years, and that public health authorities should be vigilant and take necessary action. The emergence of the 2019 novel coronavirus, in our view, confirms this prediction and highlights the urgency of taking appropriate precautions without further prevarication.[6]

This last solar cycle was at its weakest in 200 years, with an absolute solar minimum in December 2019.

This moreover underpins the fact that global warming is not currently caused by the sun itself, but by changes in the earth's atmosphere. It was none other than Rudolf Steiner, founder of anthroposophy, who on 7 April 1920, towards the end of the Spanish flu—which should actually be called the 'American flu'—pointed to a connection between pandemic respiratory diseases and solar activity. He concludes with a completely original remark that we should investigate: that this influence is intensified when the so-called outer planets Mars, Jupiter and Saturn, as seen from Earth, come particularly close to each other and close to the sun in what astronomers call a conjunction.[7] This was the case at the time of the second wave of the Spanish flu in September 1919, but it was also the case in March 2020, when Mars was seen in conjunction with Jupiter on 20 March and immediately afterwards on 31 March with Saturn, in close proximity. Likewise again in relation to Saturn and Jupiter when both planets merged—for the naked eye—into an almost single point of light on the evening of the winter solstice (northern hemisphere) on 21 December 2020. In this context, Steiner speaks of the fact that this solar activity, weakened and modified several times so to speak, can favour epidemic diseases of the respiratory organs, especially if the rhythmic interplay of head and chest organs, which expresses itself in respiratory activity, has already been weakened or disturbed before this event.[8] Just as the rhythmic activity of our chest organs is the basis of our individual life, so the solar activity in its rhythm of day and night, summer and winter with longer-wave rhythms is the basis for our life on earth. Solar activity is mediated through the earth's atmosphere in which we breathe. Due to the increasing formation of smog—think of

Wuhan or Milan as a result of industrialization—the greenhouse effect not only causes global warming, but at the same time, it also causes a certain weakening of the sun's influence on humans, largely in the big cities, where the air itself weakens the immune system in the respiratory tract, which thus contributes to severe respiratory disease. So when we ask ourselves why COVID-19 has erupted now, we can also learn to look concretely at the relationship between the Earth and the Sun, the bases of all life.

If we want to understand and address the living being of the earth in order to overcome the ecological crisis of the present, it may be necessary to develop a relationship to the sun and to the delicate sheaths of the earth as its heart and lungs. COVID-19 therefore also prompts us to question our own inner relationship to heart and breath. As we become ever more anchored in our heads in this digital age, it is vital that our life and compassion is rooted in our heart region. It is thus especially important in this crisis to imbue our thinking with feeling and not to believe that we can preserve life purely through outer control measures. A mask is not an alternative to the fact that one of the foremost tasks of generations now living is to work together for a purified atmosphere and a healthy relationship to the sun. It is worth noting that corona viruses are inactivated by sunlight.

Let us return to the document of the German Parliament that so accurately predicted the crisis and the pathogen, and note its statements on how to counter the pandemic:

> New cases are to be expected until a vaccine is available. The given scenario is based on a total period of three years with the assumption that a vaccine will be developed, released and available in sufficient quantities by the end of this period.

The paper itself points out that the virus will also change, mutate, during this time.

> Means of containment are, for example, school closures [as the first measure mentioned; precisely this has been proven ineffective as things stand today] and cancellations of major events [perhaps effective]. In addition to these measures (...) there are other recommendations that contribute to personal protection such as adherence to hygiene recommendations. The anti-epidemic measures will begin after ten patients in Germany have died from the infection. (...) the population will implement the measures differently depending on their subjective feelings. Medicines, medical devices, personal

protective equipment and disinfectants will increasingly be in demand. Since hospitals, doctors' practices and authorities are generally dependent on rapid supply, the industry will not be able to fully meet the demand and bottlenecks will arise.[9]

All in all, the current pandemic has thus far turned out to be much more merciful than this 2012 scenario envisaged. In any case, this pandemic was foreseeable and no precautions were taken! Nor were any steps taken to protect the population effectively—there were hardly any available protective masks for clinic staff at the beginning and the existing ones were stolen and sold at fanciful prices. Above all, no one had come up with the idea of specifically addressing the possible origin of this pandemic and that is the way humans and animals live together, which seems to have been a widely taboo topic until 2020—and still is.

Why was Wuhan the point of the outbreak?
There is even more to learn from the history of COVID-19. The Chinese New Year in 2020 fell on 25 January. It began a new twelve-year annual cycle with the 'Year of the Rat'. Every year, the festival triggers the largest global travel movement in the world (with 3 billion travellers in all, there were 7 million outbound trips from China and many inbound trips to China in 2019). Many of the 200 million migrant workers only see their families at this time. Probably more than 60 million children in China grow up without their parents in early childhood. Correspondingly, there are many longings and expectations associated with this festival. On 18 January, shortly before the lockdown, 40,000 families in Wuhan had already begun celebrating the approaching festival. An opulent feast with fish and chicken forms the centrepiece. This festival is also celebrated abroad, and of the 190,000 Chinese guest workers around Milan, who work in wretched conditions, many will try to go home at this time.

Wuhan itself, with its 11 million inhabitants, is one of China's larger cities at the confluence of the Yangtze and Han rivers. One sixth of all the original lakes in the city area still exist. The climate of this 'city of rivers' is humid, subtropical in summer and relatively cool in winter, similar to the Po Valley of Upper Italy. Comparable in both is the enormous degree of air pollution, not least from the steel industry there. Wuhan is the largest inland port in China and lies almost exactly in the middle between the four great metropolises of this giant empire: Beijing in the north, Guangzhou (Canton), Shenzhen and Hong Kong in the south, Shanghai in the east and

Chongqing in the west, thus forming a central traffic hub in China with its correspondingly disturbed air conditions.

Wuhan is also home to China's most important virology research institute, which conducts research at the highest security level BSL-4 [sic] and contains the largest virus bank in Asia. The USA itself funded the institute with several million dollars. The institute also conducts research of corona viruses from bats. Prominent researcher Shi Zengli found the origin of the SARS virus epidemic and publishes regularly, including on corona viruses. On 2 March 2019, she herself warned that corona viruses could jump to humans. I would like, however, to emphasize that it is very unlikely that the SARS-CoV-2 virus originated in this laboratory, as several scientists from around the world have published, for instance, in *Nature* on 17 March 2020.[10] But it is phenomenologically remarkable that the pandemic originated here in Wuhan province, although the decisive jump of the novel SARS-CoV-2 virus may even have occurred earlier in October/November.

There is a high-security wing controlling the escape of viral diseases and a meat market right next to each other; control and loss of control, almost within sight of each other. For certain—as was foreseen in the German Government's document—the market was the first site of a 'super-spreading' situation that characterizes the SARS-CoV2 virus. A few people, usually shortly before onset of symptoms of the disease and during the first days of illness, are able to infect a great many, while about 70% of those who get infected infect no one. Those doctors who pointed out this novel disease COVID-19 in the early stages were initially put under massive political pressure resulting in relevant news being withheld, as the approaching festivities were not to be disturbed. Similar patterns were to be repeated worldwide, especially in authoritarian or neo-authoritarian states. Nevertheless, looking the other way can have terrible consequences in our globalized world, and we all need to learn to reflect differently upon the global consequences of our everyday lives and actions.

In Northern Italy, 190,000 Chinese guest workers are employed mainly in the textile industry in Lombardy. In this region, the pandemic in Europe reached its first peak. This area has the highest air pollution in Europe, biodiversity is a rather foreign word around Milan. At the same time, it is Italy's richest region, intensively networked globally and a leader in digitalization as in Wuhan. On 19 February 2020, there was a football match with 2,500 fans from Spain and about 44,000 spectators. Italian politics had recently cut back intensively on hospitals. There were no masks in stock and the number of doctors was limited. Old people's homes had been privatized and gladly took in COVID-19 patients at the beginning of the pandemic. Such old people's homes are proven to be deathtraps, from Bergamo to Stockholm. The high mortality rates in Europe are closely related to the fact that many old people live in homes and that these were not protected in the beginning. It may give us pause for thought that a recent Scottish study of 300,000 people showed that the more children there were in the family, the less often adults were hospitalized and died of COVID-19.[11] COVID-19 therefore also raises the question of cohabitation and social distancing between generations.

How does this disease progress?

COVID disease is characterized by an attack on the middle realm of the human being, on the lungs, through which we permanently exchange substances with our environment, and on the blood circulation, the internal basis for our life. For infection, even the volume of our speaking and singing plays a role here affecting the amount of viral exhalation in direct contact ('the ballistics') and aerosol formation. Thus we protect others above all with the usual surgical mask, if we use it correctly, and protective eyewear can serve for self-protection. If the viral dose is low, the disease also progresses more lightly. Protection against being infected, for instance when caring for infectious patients, requires a high-quality medical protective mask, like the FFP 2 standard mask.[12] If masks are handled incorrectly, they can increase the risk of infection.

The Course of the Disease
The virus forms aerosols that can persist in the air in fine droplets for a long time. Therefore, it is recommended that rooms be ventilated regularly and that people spend more time outdoors. After absorption via the mucous membranes, infected people, who become ill, often initially notice fatigue, headaches, a scratchy throat and a dry cough, quite often. A temporary or sometimes persistent loss of taste and smell is relatively specific, possibly also diarrhoea. Fever may occur.

As early as 9 March 2020, Dr M. Girke and myself, who lead the *Medical Section* at the Goetheanum, shared a first, integrative treatment concept of Anthroposophic Medicine with our medical colleagues of the anthroposophic medical societies worldwide. Anthroposophic clinics and physicians around the world have developed their treatment concepts in close communication and have treated many COVID-19 patients at all stages of disease, as we have decades of experience in treating pneumonia without a primary bacterial cause. In July 2020, the integrative medical therapy concept of COVID-19 was published and is freely accessible online in German and English.[13]

At the end of March, at a WHO conference on treatment options of complementary medicine, there was a mutual exchange with Chinese colleagues, who at that time treated 91.5% of all COVID patients additionally with *Traditional Chinese Medicine* remedies and reported a significant reduction in the rate of severe and fatal cases of disease. It unfolded early on, that consistent treatment in the first phase of the disease is particularly important for this. In this phase, natural medicines have been shown to be effective in China, as well as in the field of Anthroposophic Medicine. Bitter substances can strengthen the organism's defences against the infection, while antipyretics and painkillers weaken it.

In this phase, the highest infectivity is shortly before and after the onset of symptoms, which can then quickly subside. The smear or PCR test (polymerase chain reaction) can detect quite specifically whether the virus is present, but it does not tell us how infectious a person is, as for instance with a child. Nor does it indicate whether and to what extent the person will become ill. In this regard, numbers of infection cases that we hear of every day say very little. The test itself is very specific when used correctly; it is unlikely that there will be many false positive tests. Test series with several hundreds of thousands of people, for example in Australia, showed very low positive test rates, sometimes as low as 0.1%. The test contains a screening method and two confirmatory tests with a highly specific gene probe. With the so-called Ct values above 30, the virus is no longer viable. There are also antibody tests, but these do not prove that one cannot infect others.

Pneumonia
Pneumonia often occurs after a good week, causing the lungs to become 'heavy' with fluid build-up in the connective tissue and in need of oxygen. Here, the use of external applications of Anthroposophic Medicine can also contribute to relief. In orthodox medicine, the drug Remdesivir has been shown to be effective in shortening the course of severe disease. Considerable organ damage in the form of 'hardening', sclerosis can remain.

Hyperinflammation and Organ Failure
Regarding hyperinflammation and organ failure: if the disease progresses, it can lead to a loss of control of the immune system, a state of hyperinflammation arises causing damage to the inner walls of the blood vessels, coagulation disorders, blood clots and multi-organ failure. In orthodox medicine, strong, fluorine-containing cortisone is used as a preventive measure to suppress such a derailment. Vital organs such as the kidneys can temporarily fail. Intensive-care medicine can succeed in saving some of the patients.

Who tends to fall severely ill and who not so much?

Age is of paramount importance for COVID-19. The low COVID mortality rate in Kenya, for example, is due not only to intensive sunlight exposure but also to the average age of the population, which is about 17 years in Africa and 47 years in Italy. The risk of death from COVID-19 at the age of up to approximately 45 years is significantly lower, or at most, as high as that from seasonal influenza, which in America (and somewhat higher in Germany) is 0.05%. By contrast, of all COVID-19 patients over 85 years, almost one in four dies on average, in medically advanced OECD countries,[14] with improvements in treatments reducing the mortality percentage of those affected compared to the first half of 2020. People with Down syndrome are a particularly severely affected group. The mortality of 40-year-olds is around ten times higher than that of people of the same age without this genetic defect. People with Down syndrome are the most at-risk group after the over 80-year-olds.[15]

The average age of COVID patients who die in Germany is 81 years. On the other hand, some 100-year-olds have survived the disease very well, perhaps also due to pre-existing partial acquired immunity to corona viruses. We can also see old age as a time when we begin to let go, when physiologically, the control of life processes in our organism also becomes weaker. On the one hand, Corona raises the question of how we relate to death in old age, because death itself, like birth, is a life necessity.

If we look at the motif of control and loss of control, this loss of self-regulation becomes significant when our blood pressure and blood sugar can no longer be healthily regulated by ourselves, when our body weight has shifted strongly in the direction of obesity, when chronic diseases, especially of the lungs themselves, weaken us. We have already listed other, social and ecological risk factors at the beginning; they are often missing from the statistics of Germany and Switzerland and can also play a significant role in other countries.

What about the children?

School and kindergarten closures were among the first measures taken worldwide, also because the narrative for COVID-19 had been written in significant parts seven years earlier and strongly oriented towards the flu. Children can significantly spread flu viruses. In this respect, however, COVID-19 behaves differently. For SARS-CoV-2, current knowledge can be summarized as follows. Children under 10 are infected less frequently, children and adolescents under 18 years of age are infected more or less as often as adults, but rarely fall ill and almost always mildly. They are admitted to hospitals even less frequently. Inflammation of the blood vessels, especially of the heart, occurs very rarely in children (MIS-C[16]). Children are often infected by adults, especially their parents.

If, like the well-known virologist Christian Drosten, we look purely at the detection of the virus and thus of infection in children, then social distancing is just as justified in children as in adults. However, this purely virological, non-medical and non-educational viewpoint neglects the fact that, especially at kindergarten and primary school age, the prevention of a SARS-CoV-2 infection has practically no health benefit for children, but it does have a high loss in developmental opportunities. However, since children also seem to infect vulnerable older adults, although only slightly, this argument has been given greater weight.

In locked-down households in South Korea, children are reported to have infected family members in 5% of cases. There is still no evidence that children in school infect teachers. The *Co-ki.de* study was able to prove only one suspected case among 9,583 children studied, of whom 194 had positive swabs and 82 had positive blood tests.[17] Children can infect classmates, and that too rather rarely. Six of 137 infected pupils who attended school infected a total of 11 classmates (Baden-Wuerttemberg). Every case in the children and adolescents was examined.

A study from Scotland cited above of 300,000 people showed that households with children were per child 10% less affected by COVID-19 cases and hospital admissions. A recent study from India of 570,000 people also concluded that although children pass the virus on to other children, they rarely pass it on to adults.[18] Contrary to *Der Spiegel's* headline, this study also provides no evidence that children infect caregivers, teachers, bus drivers or other contact persons outside the family to any relevant extent. Where we are particularly challenged by COVID-19 is our human culture of living together. The anthroposophical doctor experienced with Corona, Harald Matthes has long been calling for a differentiated protection concept for those at particularly risk. On 4 October 2020 in the *Great Barrington Declaration*, leading epidemiologists from Stanford, Harvard and Oxford University called for a transition to a protection concept that specifically protects at-risk groups, especially for older people, and calls for a far-reaching return to normality for younger people, in working life, in sport and in the area of cultural life.[19]

Such a declaration raises some concern insofar as concert-goers are often older than 55. Residents with severe physical disabilities in care homes are not only at risk at a younger age, but are also often cared for by younger people. COVID-19 has been very much about avoiding a split in society. Previous experience with so-called superspreading situations suggests a certain profile of particularly high risk of mass contagion, ranging from loud communal singing by large church congregations to lively celebrations at the Ischgl ski resort. On the other hand, there is no evidence that concert and theatre performances without alcohol, when suitable mask coverings are worn, even without 1.5m distance, lead to large infection rates. There was and is no clear evidence with regard to the real burden of disease—and not simply infection figures without regard to age and severity of disease potential—for closing kindergartens and primary schools. Effective protective measures are known to be handwashing and, yes, properly worn mouth-nose coverings can help reduce infection rates and severity in high-density infection situations. This is not something new to people in Southeast Asia.

These masks are mainly useful for protecting other people when one has already become contagious oneself. To protect oneself effectively in close contact with an infected person, professional FFP-2/3 masks are necessary.[20] There is no evidence whatsoever that wearing a mouth-nose protection in childhood contributes to the reduction of serious illness and deaths, with the exception of high-risk people, including adolescents. General mask wearing should be rejected for all children under 11 years, because the risks outweigh the benefits.[21] The main risk of increased infection can be due to incorrect use and handling.

The Question of Vaccination and a Vision of Control

The polio pandemic was an example of successful vaccination campaigns and can also show us the value of a good vaccine. Many viruses, on the other hand, such as the HIV virus, have defied every attempt at vaccine development to this day—including, until recently, vaccines against Corona viruses, in which research into vaccines for Corona viruses has not been successful for decades. For COVID-19, the goal is a vaccine that can be produced by the billions and can be easily adapted in the event of viral mutations. This is harder to do with traditional vaccines, which rely on growing viruses that are then killed or inactivated vaccines, or, attenuated or weakened live vaccines. The first vaccine developed in China made by the company Sinopharm has already administered millions of doses. The vaccine is based on such a conventional technique, which requires aluminium-based vaccine effectiveness enhancers. These so-called adjuvants are themselves a problem because they can cause undesirable side effects. Allegedly, this vaccine is effective in about 80% of those vaccinated, although scientific publications are lacking thus far. 'Officially, taking it was voluntary,' the manager of a bank in Shanghai told the *FAZ* (newspaper article on 8 October 2020). 'Practically, however, it was hardly possible to refuse participating in the vaccine trials during the test phase. This would certainly have led to a negative marking in their official personal file.' Sinopharm is officially called *China National Pharmaceutical Group* and works closely with the above-mentioned *Wuhan Institute of Biological Products / Wuhan Institute of Virology*. Here the loop closes.

Western companies in particular have developed new 'platform technologies', especially mRNA and vector vaccines, which essentially transport genetic information into the body. The body then produces a protein, for example, the spike coating of the virus, which then functions as the actual vaccine against which an immune response can then be directed. The production of the mRNA itself is cheap, but its transport into the organism is complicated. Either additives using nanotechnology are utilized, which can also trigger allergic reactions, or another virus used as a carrier is needed, a so-called vector, as with the AstraZeneca and the Russian Sputnik V vaccine. It is unclear to what extent these vaccines can trigger uncontrolled autoimmune reactions as has already been observed in isolated cases in the ongoing approval studies, for example. What we have observed since these vaccines have been used are comparatively severe initial reactions to the vaccine, especially with the mRNA vaccines, and particularly with the second dose and in younger people. Headaches, muscle pain, severe fatigue and fever can occur. A colleague of mine in alignment with good medical practice had himself vaccinated. He was exhausted on the Friday afternoon and three days later had to be operated on due to tonsillitis and a massive tonsil abscess. One has to actively 'digest' the foreign protein that these vaccines create in the organism in order to develop immunity. This process, together with its accompanying substances, can cause violent inflammatory reactions. Vital forces and rest are required. This also indicates that these vaccines can overwhelm frail old patients. Even though the safety reports from Germany's Paul Ehrlich Institute[22] state that the frequency of deaths after vaccination in very old people is in line with statistical expectations, this is only of limited comfort in individual cases when a person dies a few hours after vaccination. I would like to emphasize that this stimulating character on the organism, especially with the mRNA vaccines, is not to be evaluated negatively per se, but should be taken into consideration when deciding who should be vaccinated. A detailed account of vaccination issues for COVID-19 was published online by David Martin and Georg Soldner in 2021.[23]

What will happen after the pandemic?
In conclusion, I would like to ask to what extent today's dominant technical and economic modes of thinking and acting have contributed to this pandemic and to threats in the future from even more dangerous pandemics? Learning in this context can also mean reviewing the basic assumptions and taboos of one's own way of thinking and changing one's own attitude.

COVID-19 is *one* part of the ecological crisis of our time, which we can become more succinctly aware of through this pandemic. These days, we often feel powerless. At the same time, self-knowledge becomes ever more imperative as we recognize that we ourselves, humanly speaking, are the authors of this crisis and ultimately also the authors of this pandemic, which we ourselves predicted without ever doing anything about its triggering causes and without taking the necessary precautions. We all long for an end to this pandemic. Perhaps we should not simply long for the world as it was before the pandemic, but find out together what can lead us into the future. For if we consider this crisis in its entirety, we have to acknowledge that today life on the earth as a whole is in acute danger.

There is no breathing apparatus as such for the earth. The lungs of the earth are the forests and we humans are destroying them and with them the habitats of animals, for the sake of subjecting everything to our technical and economic control. Medicine itself has never been so technically powerful and medicine has never had so many economic resources poured into it. At the same time, the pharmaceutical and chemical industries are contributing to the endangering of the foundations of our life needs. Monocultures were only made possible by pesticides, and modern factory farming by antibiotics. Today, this threatens to make these very antibiotics ineffective for the treatment of seriously ill people, because multi-resistant bacteria are increasing. In few industries is ecology more of a dreaded word than in conventional pharmaceuticals.

COVID-19 shows us the limits of this technical and economically well-equipped medicine, which does not yet understand that a forward-looking promotion of health, individually as well as globally, needs a different scientific basis, a different thinking and way of acting, instead of the selective campaigns waged against infectious diseases and which mostly do not focus on chronic disease. If the antibiotic—which means 'anti-life'—was the most successful, even impressive medicine of the twentieth century, then the twenty-first century needs a different, probiotic, pro-life, medical approach that understands how to promote and maintain health. To achieve this, we cannot separate the health of plants, animals and humanity, both rich and poor. We need to imbue the new slogan *One Health* with life, with our feelings, with a new way of thinking and live accordingly, if we do not want to go through even more lethal pandemics in the future.

What does health mean?
Health does not come from increasingly forceful control measures from the outside, but from the ability to regulate and maintain a living balance oneself. This applies to plants, animals and humans. Health is dependent on harmonious and reciprocal relationships. If we act in such a way that we are indifferent to the destruction of the living balance of a landscape, if we are indifferent to what wild animals experience living their lives in a 40 x 60 cm cage, if we are indifferent to the fact that millions of children remain impaired for life or die from hunger, and that billions of people fall ill due to malnutrition, then we will be less and less able to help, even with the most potent of medicines.

The power of modern medicine is based on the current form of science and economics. However, it is precisely this one-sided form of science and economics that is destroying the basis of our existence, of both the earth's and human health. This pandemic is calling for a new, worldwide consciousness of what we can call the *health of our whole planet* or a real *one health*, and thus also create new foundations for a truly sustainable medicine, an ecological pharmacy. Only then can we respond adequately to a crisis such as the one we are now experiencing, which we could only predict out of our old previous ways of thinking, in the same way that we have known about our collapsing climate, but have not been able to avert it. How do we become capable of really caring for the health of the earth and the prevention of pandemics? Because this is the *real call* of the present time, which Greta's generation reminds us of, and which we must respond to together, step by step.

We are touching here on fundamental taboos of our modern, scientific way of thinking. At the core is the modern exclusion of the question of *you* in science. The question about the nature of the *other* was methodically excluded at the beginning of our modern era. Until the Middle Ages, the question was bound to God: 'Father in heaven, hallowed be *your* name'. There is no *you* in the scientific, purely material-oriented world view, there is only an *it* in scientific terms: the human being is regarded as an animal, as being purely biological; animals and plants are regarded and treated as machines, the earth as mere physical matter. This applies not only to modern natural science, but also to capitalist economics, which regards nature as an unlimited source of raw material for its own productivity, thus a mere commodity. Humanity's and the world's existence are seen as merely coincidental.

Jürgen Habermas, the great modern German philosopher, speaks of the 'methodological atheism' of the modern age, which only recognizes objects, which thinks, researches and acts in the *it*-mode of questioning.[24] If medical scientists want to research stress, they will study a mouse that, having been thrown into water, fights against drowning. It is dissected after death to detect material changes in its neural pathways, or, new-born mice are separated from their mothers in random temporal patterns to detect stress-induced, permanent genetic changes in the subsequently 'sacrificed' young mice. Furthermore, we never consider that one day what the millions of laboratory, factory-farmed and tortured wild animals have endured at our hands may come back to haunt us.

The basis of our power over nature today is the experimental freedom of researchers in dealing with nature conceived purely as an object. Our form of economy is the sister of this thinking. It revolves centrally around the increase and constant growth of money, a morally indifferent *it*, and that modern Western thinking gave the state, as its primary purpose, protection of private property. COVID-19 has shown that health is not something to be owned privately.

'Social distancing' can be useful in reducing the risk of contagion, but cannot hide the fact that my health is most sustainably promoted when I help *others* to get well. COVID-19 calls upon us to be considerate of others, and especially of those who have lived and suffered in the shadow zone of our awareness even before COVID-19. Ecologically speaking, however, all unrestricted growth leads to disease. It is known that humans and mammals are characterized by a pubescent growth spurt. This is the phase during which one breaks away from one's parents, in search of freedom, the skin develops spots and the soul aggression. This is then followed by a maturing process and taking on of responsibility follows.

Money is made by us, it is not a substitute for God, but a means of circulation to aid our economic activity. If our blood starts to grow unchecked, we become seriously ill and die, for example, due to leukaemia. Our money-based economy is a significant driver of the ecological crisis because this money lacks the necessary characteristics of maturation and death that would make it suitable as a sustainable means of circulation to begin with. The current crisis urges us to think more deeply about which form of economy would truly be sustainable and oriented towards the common good. Our planet needs this now. We are living in a partnership crisis with the earth, with the living. If we want to overcome it, we must scientifically engage in the question of the essence of the life, the essence of the earth and must enter into a living responsibility towards all that lives. Living beings are not machines and human beings are not animals. It is precisely from this that our responsibility arises. It is our moral responsibility. Whether we live up to it depends on our thinking, our compassion and above all on our approach to all that lives.

As a doctor, let me therefore briefly summarize what comes to mind from COVID-19 with regard to my own attitude: in terms of our health, we are all interdependent and cannot isolate ourselves. In this respect, we must find a common path through the crisis, leaving no one out. Protection concepts are most convincing when they include all generations in a coherent form—in recognition of our mutual interdependence. Our fundamental attitude towards plants, animals and human beings has to, in the future, increasingly be one of dialogue, incorporating respect for the life of all living things, showing compassion for the experience of all living beings and with equal respect towards the dignity of every human being. Today, we are very much in a position of meeting the basic health needs of people worldwide, and we need to transform this goal into concrete will. We can only promote health sustainably if we take the health of animals, plants and soil as seriously as our own. We need a science of the living; we need a maturing of our economy into a common-good economy. Believe me, the decisive answers to COVID-19 are not purely medical, they concern all areas of life and all of us who bear responsibility for this earth and the generations to come.

Written version of a public lecture at the Goetheanum on 12 October 2020—updated February 2021.

1. *Drucksache 17/12051—Deutscher Bundestag* vom 3. Januar 2013: 'Bericht zur Risikoanalyse im Bevölkerungsschutz 2012'. [Printed Paper 17/12051—German Parliament 3 January 2013: 'Report on Risk Analysis in Civil Protection 2012'.] http://dipbt.bundestag.de/doc/btd/17/120/1712051.pdf
2. Aberle, J. et al.: *VIRUSEPIDEMIOLOGISCHE INFORMATION, NR. 20/20. Zentrum für Virologie Univ.* Wien, 8.10.2020. (German only.)
3. See endnote 1, p. 58 ff.
4. Hardtmuth, T.: 'Die Rolle der Viren in Evolution und Medizin—Versuch einer systemischen Perspektive'. In: *Jahrbuch für Goetheanismus 2019*, Naturwissenschaftliche Sektion am Goetheanum, p. 125. (German only) [The Role of Viruses in Evolution and Medicine—An Attempt at a Systemic Perspective.]
5. Jiangwen Qu und Chandra Wickramasinghe: 'SARS, MERS and the sunspot cycle', *Current Science*, Vol. 113 (8), 2017, 1501–1502. Accessed online on 25.06.21. https://dunapress.org/wp-content/uploads/2020/03/Sars_MersAndSunpots.pdf
6. Cf. 'Space weather and pandemic warnings?' March 2020, *Current Science*, Vol. 117 (10), 1554. Accessed online on 25 June 21, at: https://wwwops.currentscience.ac.in/Volumes/117/10/1554.pdf
7. Steiner, R.: *Introducing Anthroposophical Medicine.* (CW312) Lecture 18. Published: SteinerBooks (October 1, 2010).
8. Ibid.
9. See endnote 1, p. 73.
10. Andersen, K. G., Rambaut, A., Lipkin, W. I. et al: 'The proximal origin of SARS-CoV-2';. *Nature Medicine 26*, 450–452(2020). https://doi.org/10.1038/s41591-020-0820-9
11. Wood, R. et al.: 'Sharing a household with children and risk of COVID-19: a study of over 300,000 adults living in healthcare worker households in Scotland'. *BMJ.*

12. Hemmer, C. J. et al.: 'Protection from COVID-19: The Efficacy of Face Masks.' Meta-Analysis. *Dtsch Arztebl Int.* 2021 Feb. 5; DOI: 10.3238/arztebl.m2021.0119
13. Georg Soldner, Thomas Breitkreuz: COVID-19 on *Anthromedics*. https://www.anthromedics.org/PRA-0939-EN
14. See endnote 2.
15. Robert-Koch-Institut, *Epidemiologisches Bulletin 2/2021*, p. 36. (German.)
16. MIS-C—Multisystem Inflammatory Syndrome in Children.
17. Schwarz, S., Martin, D. et al.: 'Corona in Children: the Co-Ki Study: Relevance of SARS-CoV-2 in outpatient pediatric services in Germany '*Mschr. f. Kinderheilk.* 169,39–45 (2021) https://link.springer.com/article/10.1007/s00112-020-01050-3
18. Laxminarayan, R. et al.: 'Epidemiology and transmission dynamics of Covid19 in two Indian states'. *Science* 06 Nov 2020. https://science.sciencemag.org/content/370/6517/691
19. Dr M. Kulldorff, Dr S. Gupta, Dr J. Bhattacharya: *Great Barrington Declaration*. https://gbdeclaration.org/
20. See endnote 12.
21. Ines Kappstein, *Krankenhaushygiene, up2date 2020*; 15(03): 279–295. (German article.) [Mouth-nose protection in public: No evidence of effectiveness.] https://www.thieme-connect.com/products/ejournals/html/10.1055/a-1174-6591#info
22. Paul-Ehrlich-Institut: Sicherheitsbericht, Verdachtsfälle von Nebenwirkungen und Impfkomplikationen nach Impfung zum Schutz vor Covid19 seit Beginn der Impfkampagne am 27.12.2020 bis zum 26.02.2021. [Safety report, suspected cases of adverse reactions and vaccine complications after vaccination to protect against Covid19 since the start of the vaccination campaign on 27/12/2020 to 26/02/2021.] https://www.pei.de/SharedDocs/Downloads/DE/newsroom/dossiers/sicherheitsberichte/sicherheitsbericht-27-12-bis-26-02-21.pdf?__blob=publicationFile&v=9
23. Martin, D., Soldner, G.: 'COVID-19 Vaccination: A synthesis review of current status and proposal of a registry study to overcome social polarization tendencies and answer open research questions'. https://www.anthromedics.org/PRA-0971-EN. (Updated 05.03.2021.)
24. Habermas, J.: 'This too is a history of philosophy.' *Vol. 2, Rational Liberty. Traces of the Discourse on Faith and Knowledge.* Suhrkamp/Insel, Frankfurt/M., 2019, p. 191–211. https://www.suhrkamp.de/buecher/also_a_history_of_philosophy-juergen_habermas_58734.html?d_view=english

Gerald Häfner

We Poor Royal Children:
Corona and the Social Challenges of Our Time

Humanity in Crisis

We are currently living in a state of emergency. It is a paradoxical state because we are behaving completely paradoxically within it. For if, from a social and a medical point of view, one were to ask, 'What actually keeps me healthy?', then the answer would be that I go out into nature, that I am active, that I have encounters, that I experience meaning and emotional warmth, experience touch, recognition and appreciation. However, at present we are constantly required to behave in exactly the opposite way, which I feel is not really appropriate in the situation, nor very human. For we move out of each other's way when someone approaches us; we don't hug each other, but keep our distance. We wear masks and hide the part of our face that is most alive when we speak. A really strange state of affairs! I very much wish that we don't get used to it.

In this article, however, I am not addressing the day-to-day political minutiae of this issue. It is of course important, but not at this point. Rather, I would like to try to reflect on this together with you: What is the *signature* of this current state of affairs and above all, what is at stake? What is it all about? Moreover, what can we learn at this time and how can we nevertheless shape it positively?

My impression is that we are living in a time of multiple crises, one after the other. One crisis is immediately followed by the next. There is the fundamental social crisis of capitalism, which is expressed by the fact that, depending on how you calculate it, 8 to 32 people on this planet own—in monetary terms—as much as the entire poorer half of humanity, amounting to about 3.7 billion people all together. This is unacceptable! This is really a serious crisis and it has not been overcome. Then there is the financial markets' crisis, which has preoccupied us since about 2008 and has turned into the so-called euro crisis; a crisis that has led to the fact that the public sector, being human communities, municipalities, countries and states, has become increasingly indebted, while at the same time, private wealth is growing continuously to an ever greater extent, and as it does so, the fewer people there are at the top of the list. This means that the greatest wealth is in the hands of a tiny number of people.

Then there is the ecological crisis, affecting all of us for at least half a century, in which we are poisoning the oceans, poisoning the soil, poisoning the air and by 2050, there will be more plastic floating in the oceans than living creatures, if we continue as we have done up until today. None of these crises have been solved. To top it off, there is the climate crisis, which has really shown its grim reality all over the world during the last year, in drought, in dying forests, in fires, as in the Pantanal in Brazil, in Australia, in California, in Siberia—in fact everywhere on earth!

These crises have not been overcome. That is why I have to correct something I said earlier, because the reality is that we are not dealing with successive crises, but rather with overlapping crises that humanity is facing at the moment. None of them have been solved! All of them are still with us, none of them has a solution in sight—and more are still to follow.

Now, since the beginning of 2020, there has been almost no other topic of conversation than the Corona crisis, this pandemic! Even before this with regard to the climate crisis—which of course continues—many said, 'The earth has a fever! We have to protect our earth! Shouldn't we finally start changing our economy and lifestyles? From now onwards, couldn't we behave differently towards our planet, towards plants, towards animals and towards each other, so that we will still have a future on this earth and not ruin everything?' Pupils didn't go to school on Fridays so that they could organize school strikes for highlighting the climate crisis. How serious do they have to be for us to take such actions seriously? How unbearably little has actually happened since then. How much have we simply continued in our old ways with our everyday habitual lifestyles? How quickly we said, 'Yes, you are basically right! But we can't change so easily and, above all, change so quickly. We will have to learn how and then get used to it!' And then a virus broke out in Wuhan, in China—and suddenly it was not the earth, but people who were running a fever! Everything was different and everything had to change, and it happened suddenly and quickly.

Where do we stand today?

For me, this is the first important consequence of the Corona crisis—we can, if we want to, change everything and quickly! It is in our hands! We always have the freedom to act. We have no excuses!

The second thing is that these crises—all of them and the current pandemic in particular—are crises that affect humanity, all of humanity, no one is exempt. We are all in this crisis together, even if we are affected differently. I have the impression that this crisis is something like a last warning signal placed before us. Not simply 'Keep it up!' rather, 'Change your thinking! Start thinking differently, feeling differently and living differently!'

I would now specifically like to look at this situation more closely from a certain perspective. Of course, this is only one perspective, and not the only one.

Where do we actually stand today? We have, as humanity, come a very long way. It began with an early state—which incidentally, all cultures, all religions describe similarly—a state of all-encompassing unity or total-connectedness. Christianity describes this through the image of Paradise; in other cultures we find other descriptions of it.

What is meant, however, is a state in which we human beings were basically not yet really separated from each other and also not separated from what is outside of us from today's perspective: from the trees, the plants, from all beings that live around us and are connected to us. The mountain was not just a pile of stones, but was animate and spiritual. The river was inhabited by river spirits, the tree was inhabited by spirit beings. When people cut down a tree, it wasn't just a piece of wood fit for a chain saw, but you first spoke to and asked the tree being, and if you were allowed, you cut it down. The tree became a boat or a house. The tree was given a meaning, a context. That was our relationship to the world. I have had the great good fortune of visiting groups of people several times in my life who still live in this consciousness. It is an immensely moving experience.

However, it was also necessary that we human beings—at least the majority of us—left this state. The Bible describes this with the breaking of the one single commandment that existed at that time, which was, 'Do not eat from the Tree of Knowledge!' But that is exactly what the first people did; they broke the rule, ate from the Tree of Knowledge and in so doing left the paradisal state.

This created a division between us and the world. The oneness had split. Humanity had broken away and so began the separation of subject from object, in which we stand now disconnected from the world around us. As a result, we see the world as something that is available to and usable by us at will. The world appears as something we can own, control, understand, penetrate, exploit and use for our own purposes, while at the same time, increasingly losing our connection to it. Today, as Oscar Wilde once wrote, we live in a world in which we 'know the price of everything and the value of nothing'.[1] We have learnt to regard the world within a certain framework, one of numbers, measurements and weights. It is abstract quantities that are important to us, for instance, prices. How much does a kilo of wheat cost on the market today? What will it be tomorrow? Can I make a bet on the price trend and the fraction of a percentage profit from the increase in value (or vice versa from the decrease in value), if I predict the trend correctly in advance. Will I make a fortune, if I just make a clever move or the computer regulates it for me via an algorithm?

That is the world in which we are living. A world in which real things, for instance all that nourishes me, have increasingly lost their meaning and significance, and where we conduct business solely with the corpses of plants, animals and the world as a whole, as it were. A world where we look at everything as if through the lens of the dead—to all that is quantifiable and measurable, only the material. We have chosen this path and have consequently set up a mechanistic world for ourselves, a kind of functional machine. We have also pervaded the world with rules that determine our behaviour and lead to the fact that we often no longer ask what is right, what is appropriate, but only ask what is my right and how can I gain, profit and earn the most out of it? This lives so strongly in our consciousness that it has become the signature of our times. We have been brought up and learnt about it in this way. We have constantly, actually daily, been trained by circumstances to always only look out for our own self-interests.

This is the attitude with which we encounter each other and the world, and this is also the starting point of the economic order that determines our world today. It assumes that the more each individual pursues his or her own self-interests, the greater the good for the community, and vice versa. Adam Smith himself initially called this observation 'the invisible hand'. Somehow this function was puzzling to him; he did not fully understand how it actually worked—'that everyone thinks only of themselves, of their egotism and from this arises the greatest possible benefit for all?'. Later he simply called it the 'economic law'. It still forms the basis of classical economic theory. This has unfolded in such a way that we are endlessly calculating. Economics has become more and more only about mathematics. It increasingly obscures what is being calculated. Neither machines nor devices actually run the economy. People run the economy. However, in order to be able to compute people, we first have to make them 'calculable'. To do this, economics had to create a definition of the human being, the *homo economicus*—a rational person who continuously pursues increasing wealth for his own self-benefit.

This is what we humans have become, from this perspective! A being that continually wants to increase its own benefits by using its rational mind! At the same time, circumstances reveal to us that if we continue in this way, we will lose our connection to everything of real value in the world, our connection to the earth and our connection to life. This includes our connection to other people, who are becoming more and more objects to us and with whom we no longer truly engage in relationship.

We notice very clearly—and I think all these crises show, at their core, the same thing—that we cannot continue like this. We have to become able to reconnect, to enter into relationships, to live in resonance with the world and our fellow human beings. Actually, if we think about it further, we have to find this connectedness and oneness with the world again, of course in a new way, but not by going backward. Going back is not possible.

This is both the tragedy and the blessing of human development, that the way back always gets blocked. If we want to go back to earlier states, we will end up in evil. In the twentieth century, when situations were constructed that mimicked the ancient Egyptian empire, pyramidal with a pharaoh at the top and the people at the bottom, we ended up in evil, as with Hitler, Mao, Stalin, Pol Pot, to give just a few examples of where this leads. That time is over. We live in a different time today, in a different state of consciousness, and the question is whether we are also capable of developing social conditions in such a way that they correspond to the state of consciousness we have attained. For from this arises a drama that is as fascinating as it is frightening, one in which we humans have long since moved much further ahead in terms of our inner consciousness than have the conditions in which we live.

Making conditions human

There is a survey that has been conducted every two years in Germany over many decades—the so-called 'system satisfaction' survey. People are asked, 'How satisfied are you with the system in which you live?' One of the questions is always, 'How satisfied are you with the economic system in which you live?' In the last survey, 90% said, 'I am not satisfied with the economic system I live in.' Ninety per cent! Nine out of ten people in Germany are not satisfied with the economic system. And when they were asked what their reasons were, people above all said, 'It is not just! It is not fair. It leaves us feeling inwardly cold. It ensures that we can no longer trust each other, but in fact have to mistrust each other. It actually undermines everything that is supposed to be economic, which is above all the cooperation between people.' So we have set up something that nobody really wants. We continue acting according to laws that nobody really wants any more. Actually, we have made some progress, but the big question is whether we will finally be able to transform the conditions in which we live.

Today, we need to shape conditions in such a way that they become more human—and that means that we become more aware of our responsibility towards our fellow human beings and posterity. This is the great social question of our time. It follows on from the medical question in Georg Soldner's contribution in the previous chapter. He ended his contribution with the social question, in which he considered the core of what is currently happening and the solutions to what is currently happening. Just as in the case of a disease that affects an individual person, we can also look at the various factors and elements that play a role in it and affect the social sphere.

First of all, there are always so-called triggering events. This may be a virus, for example. However, there is also the question of the strength of each person's immune system—how strong is the person's ability to deal with the triggering cause? How does he or she, how does his or her organism deal with this virus? It is the same in the social sphere. On the one hand, there is the external event such as the one that we are constantly staring at, at the moment. We are captivated by this tiny issue, the virus—particularly, how this virus can be avoided. On the other, the question is how we deal with it in society, and whether we have or can develop the necessary immune response, figuratively speaking.

Incidentally, this time also offers an interesting social learning experience. I, for example, am learning a lot at the moment in new ways of greeting people. For decades, I simply greeted people in a certain way. I usually shook their hand and squeezed it more or less tightly. Today, I don't do that automatically anymore, but pay attention to whether or not it's right in this case and with this person, in this situation and at this moment. There is no one rule defining what is right and what is not, but it needs to constantly arise anew from the awareness of what is appropriate between us right now. In the process, a force grows within us that, when it becomes stronger, can increasingly replace external conventions. These external conventions are always only crutches for what actually needs to happen as an internal process.

Outwardly, you can never really say what is right. This also applies to the current regulations. Why six people gathered in a room is acceptable, or too many, or why shops have to be closed if 35 in 100,000 people in the population have tested positive. These are all arbitrary variables. The question of what is appropriate to protect the other and feel safe is at least as much about the other person as it is about myself. This question is actually something that is constantly arising between people, as if it were new. We have already reached a level at which we can experience and practise what is actually required and needed by us in the interpersonal and beyond.

It is significant that this virus primarily affects the respiratory organs. After all, breathing is what really connects us to each other, more than anything else. We actually breathe together. The air I breathe out is the air you breathe in, and vice versa—and this extends across the entire world. The breath also not only connects us to each other, but it also connects us in a very particular way to plants. What the plants breathe out, we breathe in—and vice versa. It is a constant give and take, a harmony. The whole of life is wonderfully harmonious, if we understand it this way. Ironically, we are nonetheless in the process of destroying this harmony.

This pandemic has a lot to do with how we treat each other and how we treat our fellow human beings, our environment and posterity.

What do we need to learn anew?

It seems to me that it is the same for the individual human beings—every illness is not only an external but always an internal occurrence as well. It raises the questions for me—Why me? Why now? Why this? Actually it is always the case that I can understand—at least with the passage of time—what the illness is trying to say to me, and this helps me to develop new strengths and new abilities. In the same sense, I believe that we, as humanity, should learn and practise asking in such a pandemic situation: What does this all mean for the biography of humanity? What do we want to learn from it now? What is this time telling us? What are our tasks?

It seems to me that what needs to be learnt has to do with the fact that, in the meantime, we have all become relatively good at taking responsibility for our individual lives. Everyone has learnt this. We can achieve this to some extent. However, what we have not yet learnt is to take responsibility for our lives together on the earth. Let me put it another way. We are all conscious of what we mean when we say 'I'.

There is an 'I' in us that like a conductor raises the baton. I can give into hunger or greed, lust or tiredness. Alternatively, I can also decide, 'No, I'm going to finish this essay and eat something later'. That is the 'I'! Something that suddenly intervenes and says—here I feel this, there that, here yet another preference or inclination. Nevertheless in this house, in this orchestra, I am in charge and I decide what I want to connect with now, what I want to do, which path I want to take, which melody I want to play.

Where, on the other hand, is this 'I' in society? Does it even exist at all? Who is setting the course? Who determines what happens? How are interpersonal relationships in society today?

There are some of us always very quick in thinking we know exactly why and how everything is and how it should be. This is always dangerous. These are the people I am most afraid of. If anyone in the world today points to a certain person at a certain place and thinks that he or she caused all of this and should now answer for it, then this assertion is in fact already wrong! Because the world and reality is never that simple. It has an infinite number of layers.

What is required today instead is that I not only take responsibility for my own life, but that we all become aware that we also bear responsibility for others, for the whole. Thus, an 'I' is gradually emerging in society, one we are forming by being human beings together—this is very different than in the past.

In the primordial history I spoke to you about earlier, the need for all-encompassing unity did not even exist. Later, in the first forms of human culture, in the early human groups, we find ways of living together gradually evolve. I will only sketch this out very roughly. We find a stage of development where groups of people were governed by leaders. This went on for a very long time and only changed in character very slowly. In Egypt it was the pharaoh, in Russia the tsar, here in Europe, the king or the emperor. For a very long time, there were always people who determined from the top down what happened in society. As a citizen, it was comparatively easy because you didn't have to decide anything. However, it was also tragic because you were simply subject to someone else's will. You were really just an object of their decisions. That time is now over.

Today we are living—anthroposophically speaking—in the age of the consciousness soul and the consequence of this is that we are living in an age in which what happens socially, what is socially right, can no longer be determined by one person, but can only result through the collaboration of everyone. This is a completely new ability and one which is needed today.

Furthermore, for those who are concerned with anthroposophy and the threefold social organism, this makes it understandable why Rudolf Steiner spoke of the necessity for 'self-administration' in 1919.

We are often only aware of the one aspect of social threefolding, the distinction between spiritual life, legal life and economic life; the functional division and the qualitative allocation of freedom in spiritual life, equality in legal life and fraternity in economic life. However, what is inseparable from this, and what Rudolf Steiner tirelessly points out, is that above all self-management in spiritual life, self-management in legal life and self-management in economic life has to be developed. It follows that things should no longer be decided from the top down—which by the way is still the case today regarding what is to happen and what is not to happen—but that the people themselves now take responsibility. In 1919, self-determination was something that no one really knew, had experienced, let alone practised. Rather, people had grown up in an empire and were accustomed to the fact that orders were directed from above. So Rudolf Steiner's descriptions met a population that was immensely interested, but not yet capable of bringing into reality something that was needed, already at that time.

Today, 100 years later, it is really time for us to develop the ability to take hold of this kind of self-determination, for if we do not, it might be too late. We are in a situation in human history where it is a matter of crossing a threshold in the acquiring of new capabilities with a whole new quality, which is the quality of creating an 'I' together with others in the social organism. Even if this irritates you, I believe the pandemic, in which we are currently living, is a setting in which we can begin to put this into practice. We just have to learn to refrain as much as possible from the fixation on ourselves. Or to put it another and better way—to include the world context in our consciousness, feelings and actions more and more.

I was of course grateful that I was able to experience the lockdowns in Dornach in wonderful natural surroundings, with the possibilities of going into the forest, climbing the mountain and such things. Nonetheless, I am also aware what a lockdown means, for example for people who cannot do that, who have to live in slums, in overcrowded big cities, where they are crammed into the smallest of spaces and where there is no way of going to the countryside, and are generally unable to move freely. We have to be aware of all of this at the moment.

Important steps in consciousness

It is also good to have in our consciousness what it means when families cannot work, when children are no longer allowed to go to school, or the effect on countless people who have currently lost their jobs and their incomes—because of Corona. We need to hold in our consciousness what it means socially when, for example, the number of acutely hungry people doubled in the first half of 2020. I don't want to avoid providing another relevant statistic. Parallel to the doubling of the number of hungry people, deposits at Swiss banks have climbed exorbitantly. For example, at the Cantonal Bank of Zurich alone, deposits increased by 444 % within six months. All this is part of today's social reality. It affects people differently. The poor are in acute need and the wealthy are worried about their prosperity. So we can be aware of how people in other countries become scared when they realize that things are getting socially tight and that, at the moment, governments are spending enormous amounts of money on protecting and saving people. It could happen, the wealthy worry, that governments will want to take money from them through taxes or levies. These people therefore move their money to Switzerland or other safe places. Switzerland is a country where 23% of the world's flight capital is invested. I say this because it concerns all of us that we become conscious of where we are, how the world is set up, and what is profited from here.

Clearly, the social question is far greater today. It no longer only concerns interpersonal relationships. I would therefore like to address yet another layer. This virus has in all probability jumped from an animal species to humans—just as, incidentally, all great human epidemics were originally zoonoses. What has been researched very well in recent times is the connection between the emergence of such zoonoses and our dealings with the animal world. If we look at Ebola in Africa, for example, the virus outbreaks that occurred were almost always in areas where virgin forest had previously been cleared, where countless animal species had been deprived of their habitat. As a result, they were forced, under stress, to flee to other areas where they could no longer live naturally. This is how we first create the conditions in which microbes, viruses that have lived peacefully and without problems for the animals themselves for ages, mutate under stress and jump over.

By the way, it was the same with the HIV virus, which was transmitted from macaques to humans. Exactly there, where we constrict, where we destroy habitats, where we deforest large areas, placing the animals under stress and where they can no longer live their natural lives, such events occur. It is similarly dangerous when wild animals are crammed into confined spaces, as was the case for example at this wet market in Wuhan. I have seen several such markets myself. It is a spooky world. How cruelly live animals are treated there. They are also slaughtered next to caged animals and in front of the people who buy them. Fear and unbearable 'social' stress for these animals strongly promotes conditions for the emergence and spread of zoonoses.

A major paper on this appeared in *The Scientist* in January 2019. A United Nations programme studied this connection for twelve years and developed proposals to prevent future pandemics. This is the so-called *One Health Programme*. This research programme ran on behalf of the United Nations by a research community that researches these connections worldwide under the title 'Predict' and has its headquarters in the USA. Under President Trump, however, the USA demanded that this programme be discontinued and refused any cooperation or support. Previously, these researchers had pointed out in no uncertain terms that if the plans of former President Donald Trump in the USA and Jair Bolsonaro in Brazil were to become reality, we would experience completely different, far more dramatic pandemics than this current one as a consequence. These plans included freeing industry of all environmental regulations, allowing deforestation of large areas—where real nature still lives and reigns—or, as Bolsonaro put it, squeezing the tropical rainforests like a lemon. If carried out, thousands of animal species, some of which are not even yet known, would be deprived of their habitats.

Therefore, if you want to do something about pandemics, it is not simply a matter of putting on masks. That is, so to speak, only the outermost, most superficial layer of events. If we want to do something about it, humanity has to, for example, change its relationship to the animal world and likewise to the plant world. It is the opportunity and the task of the world community, the human community, to finally wake up and recognize what is at stake and that it is high time.

It is about time

It is about time that we break through the barrier to consciousness of what we have created for ourselves by always remaining on the surface of the materially visible and by not being able to penetrate to the essence. It is this outer, materialistic object consciousness that does not allow us to penetrate to the essence of ourselves, to the essence of other human beings and to the greater, comprehensive essence of the living and the organism of the earth.

Anthroposophy considers itself to be a path that helps each human being to achieve this breakthrough. I do not turn back to an old atavistic consciousness, but practise orienting myself with my modern, clear, discerning and scientifically-trained consciousness, in such a way that I become capable of recognizing not only the dead, but also the living and thus the future, the becoming; that I become able to encounter 'beingness', that I also become capable of relationships with the realm of the living, with all beings around me. This is anthroposophy!

Anthroposophy is not only knowledge. Nor is it only self-development. Anthroposophy is also practice—and here this means, the commitment to a better world. In the watershed years of 1919/1920, Rudolf Steiner, founder of anthroposophy, himself eventually became something of an activist, taking to the streets and to employees' meetings and getting involved in the social debates of his time in Baden-Württemberg. For it was clear to him that there would be no future for humanity unless we change our social conditions, transform the social organism and transform it in such a way that freedom, equality and fraternity are manifested at all levels of society. Today, more than ever, it is our abiding task to finally take responsibility for the social organism, which is not a mechanism, but a living being that we shape and determine ourselves. Only we can shape this social organism appropriately—and only together. This task is before us today and the crises, especially now the current pandemic, can point out to us that we have to make this breakthrough at all levels, which I have just tried to describe, and in the field of the economy as well.

It is necessary to make the breakthrough from a science of counting, measuring, weighing and systematizing—from a science that, so to speak, only recognizes the dead with certainty, but does not yet have adequate methods and concepts for the living, for beingness—towards a genuine science of the living. A science that does not just stare at a single point, but takes into account the surroundings, the context; a science that does not isolate, but makes us capable of relationship and encounter.

In the field of politics, we have to arrive at fundamentally new forms of a social order where power is not exerted from the top down, but is shaped by the self-determination of the people—from heteronomy to self-determination. We need to move from a spectator democracy to a direct or participatory democracy, step by step. We no longer need top-down politics, but bottom-up politics. The major crises of the last 100 years, in particular, have shown that countries that are resilient and successful tend to be those in which there is a high degree of responsibility for one another. They are also more likely to be countries with a lively, active civil society. Interestingly, it is also countries often governed by women that have been more successful in this pandemic than those governed by men. A dialogue-style of politics is proving to be more sustainable and successful than a unilateral one. The worst performers were the great narcissists who for a long time first denied reality and then blamed others, and divided the country.

We realize that certain ways of thinking, feeling and behaving are simply over and are no longer effective at this time; that it is a matter of developing other, new cultural solutions. This is a task that we all have to face together.

In the economic sphere, this means moving from a competitive to a cooperative economy. It means moving away from an economy in which the next person is my competitor and I have to try to keep him down or outdo him, to an economy in which we work with and for each other. This new outlook is—what can I do for you and all of you, and conversely, while all of you are doing something for me at the same time.

This is what the anthroposophical social impulse is striving for: full freedom in spiritual life, that is, a truly free cultural life in the spiritual sphere, and in science, independent of state and economic influences. A culture that reflects the spiritual self, the limitless possibilities of knowledge and does not submit to a form of reductionist and exclusively application-oriented servitude of the spirit.

In addition, a culture of rights in which—as Rudolf Steiner puts it—from now onwards, we can only regard as right, laws in which everyone has had the opportunity to participate. Laws arise between people. It is not the established, given laws that are important, but the equal formation of new laws from within the heart of society.

Finally, in the economy, the cooperation of all with and for each other is the working principle. In reality, we have been working for other people for a long time and not for ourselves, just as other people have been working tirelessly so that we can exist. Fraternity has already been established, but is not yet fully understood. That is why we are still living within old forms.

What Rudolf Steiner described at that time under the heading 'the threefold structure of the social organism' is not an ideology. It is not an abstract concept, but an attempt to describe—as Goethe did with his example of the 'archetypal plant'—what operates as the basic social forces in every society and in every human being. Freedom, equality, and brotherhood is as fundamental within the social domain as thinking, feeling and will lives within every human being. Rudolf Steiner also describes how society has to change in order for these basic principles to take effect. At present, for example, fraternity is not brought to bear in the economy because it is set up in such a way that I am constantly forced to think only of myself and to view the other person as my competitor. If companies want to work in close cooperation, it is often even prohibited, for instance by anti-trust laws.

It is similar in legal life. Here too, we are far from having a culture in which we form and shape laws together. Rather, we are subjected to laws shaped by others and from above. Just imagine for a moment how different things could have been in this pandemic if authorities, such as the governments in Bern, Berlin, Paris or Vienna, did not uniformly prescribe what everyone should and should not do, but handed this task to the self-management of the different branches of society. Let us take the school system as an example. If schools were not simply closed for four months via a centralized decision, but were rather organized in round tables to whom it was said, 'Dear schools, we now have this and that problem and let us explain the important factors to you.' Then ask whether, through the self-administrative bodies of schools and the school system, they could develop sensible educational ways of working, while at the same time, taking these factors into consideration. I am sure that completely different and much more meaningful and healthier ways would have emerged than those that were ordered from above through state directives.

I believe it takes a certain amount of mature civic-mindedness today to do three things at the same time, in such tense times: that is to clearly criticize what is wrong, to address very precisely what needs to change, while at the same time, adhering to what needs to be adhered to, so that society can hold together and go through such an ordeal together.

I experienced this time-out in spring, when suddenly everything stopped, twice; a cultural and a social catastrophe. At the same time however, it also gave an opportunity for contemplation, reflection and a new social beginning, an occasion to ask myself anew what really is important? What is important now? Do we want to go back to business as usual afterwards? Or not? And how? We are still in the middle of this situation. Let us use this time to not continuously dwell on what disturbs or does not disturb us right now in our everyday lives, but to approach what is necessary within ourselves, what is necessary between each other, what is necessary in our community, what is necessary on a global scale, so that as humankind, we can further develop, take responsibility for and lead this earth, life and ourselves into the future in a meaningful, good and healthy way.

The new royal children

Let me now conclude by putting what has been said into a very succinct picture once again: there have always been kings. Now they are gone. But we still behave as if kings exist, as if someone else is responsible and not ourselves. We still have the feeling that we are not responsible for these big questions that have been mentioned in this article. Sure, we can talk about it or rant about it, but the responsibility for them lies with others, politicians and governments, or maybe even business leaders. We don't really care. We leave it to them. That is, we behave as if there were still a wise king who orders everything from above. If we glance upwards to where the decisions are made, we can see that there is no king there any longer. Wisdom has gone. It does not exist in those elite places. It is as if wisdom has died there, but has been reborn in each individual human being according to his or her own potentiality.

The royal might that used to lead people from the outside, from above, is now reborn from below in each one of us, as an inner 'royal' force. So we poor royal children, abandoned by external leadership, must begin to lead this world anew from within. The crown of creation, humanity, is in the process of destroying that very creation. However, we can now crown each other; crowned human beings for whom the earth and creation are waiting. For I am absolutely certain that we human beings have a task on this earth, not only for ourselves, but for the plants and animals as well; a task for the royal children for whom this world is so eagerly waiting.

Slightly revised written version of a public lecture at the Goetheanum on 19 October 2020.

[1] Oscar Wilde, *Lady Windermeres Fächer* (1892), 3. Akt.

Jean-Michel Florin

How Is Our Behaviour Mirrored In The Ecosystem?

Prevailing Perspectives in Agriculture

How can we see the reflection of our behaviour in the ecosystem? Is it observable at all? Is it relevant? I find this an unusual but interesting question and will begin with a small illustration.

I live in France, travel by train and often walk past the many allotment gardens as I wend my way home. Each garden looks different. I don't know all the people, but I have a clear impression of the relationship between the garden and the individual person, because I can see how the gardens have been worked, throughout the year. There is a very perceptible mirroring of the people who tend them and their respective world views. It is actually a little funny, for example, because there are two gardens next to each other that have garden gnomes, plastic storks and all manner of ornaments like this. They are very similar, the two gardens, as if they have somehow imitated each other. I would need to talk to the people one day to find out what relationship exists between them and their two gardens.

With this I want to demonstrate that when we work in nature, shape it and interact with it, we actually create it the way we ourselves view and understand it. This is evident everywhere and can be proven. Everything that lives inside us as an image, as a world view, ultimately flows into our actions. Therefore, if we now turn our gaze to look at the world and ask ourselves, 'How do we actually see nature? What is our view of nature when it looks the way it does at present?' It may somewhat frighten us.

I would like to try to develop the view of nature that prevails today out of the current state of affairs and the current Covid crisis. What does nature, which today is shaped everywhere by humans, actually say about us and our inner attitude towards it, our world view in the truest sense of the word? If many people succeed in taking this increasingly seriously or at least become aware of it, then something will change, because if today everything were wonderful and the world and nature were a paradise, then not so many people would be concerned about it.

Insights Gained Through the Crisis Situation

We are undoubtedly in a crisis situation. What does crisis mean or signify? From the Greek word '*krisis*', crisis means: problem, abyss—yet it also means a resolution towards something new. We should take this seriously! There is indeed a crisis, there are indeed problems. However, at the same time it is a moment in which we have the opportunity to really take hold of the problem as a crisis, and to resolve to decide for something new. In this sense, the current virus is a major crisis for the whole Earth, but one that perhaps also contains positive opportunities for the future.

This virus has surprisingly kept the whole world in suspension, holding its breath—that's fair to say. On the other hand—because I have a lot to do with farmers—from an agricultural perspective, it is not such a big surprise! In agriculture, we have had new epidemics, a new virus, a new bacterium or a new fungal disease every year for more than 30 years, on a regular basis! I could fill a whole hour with such epidemics alone. This year, for example, noticed only by a few, there's been an unpleasant tomato virus. This has been terrible for many tomato growers, especially those who specialize in tomatoes. For others who prefer diversity in vegetable cultivation, and also grow tomatoes, this particular virus has been less of a problem. However, the virus is serious and the tomatoes are inedible. There is also a swine flu, which has been steadily circulating this year and has just crossed the border into Germany. French customs officials are already highly vigilant, wanting to stop it at the border at all costs. Year after year, there are ever more such devastating viral, bacterial and fungal diseases; in plants, in animals—and now also in humans. What does it really indicate that we've had a significant increase in these viral and bacterial diseases in agriculture for the last 30 to 40 years?

When applied to the whole of nature and human beings, this means that there are actually bigger problems than just COVID-19. The Coronavirus only sheds a different light on this greater crisis. While the number of diseases is increasing every year, biodiversity is decreasing massively. There is a clear relationship between this decreasing biodiversity and increasing bacterial, viral and fungal diseases. Why is this so? This has already been investigated quite well in part, and it's been found conclusively that the emergence of many new viral and bacterial diseases is linked to the destruction of entire ecosystems.[1]

What is an ecosystem? It is simply a particular location; it may be a particular forest or meadow, where a variety of microorganisms in the soil, plants and animals live together symbiotically. In addition, there are several hundreds or even thousands of species of flora and fauna. Even in a sparse Scottish grassland, there is more biodiversity than in the Amazon forest. Of course, the plants and animal species are much smaller there than in the rainforest. In any such ecosystem, a great many creatures live together forming a much larger kind of organism. Not just eating and being eaten prevails there. It is much more complex! There are many interrelationships and regulating processes taking place.

Let us imagine, perhaps, such a diverse, beautiful meadow as an inner picture—around 7,000 microorganisms, flowers, animals and bounteous insects living together there. If a hole is made in the meadow, a place where plants have been removed or are missing, what happens then? Weeds multiply very quickly, and it won't be long before the bare spot is overgrown again. If the hole in the meadow or ecosystem is very large, then a so-called 'weed' will quickly occupy the whole spot, crowding out all other plants, which will gradually disappear. This 'weed' becomes stronger and stronger and takes over.

This is roughly what happens when an epidemic occurs. The viruses or bacteria are usually integrated into very specific natural networks or cycles within many different species, such as the coronaviruses currently living in bats. In nature, regulatory systems take over when gaps or 'wounds' occur in the ecosystem. If the wound is too big, if the ecosystem is fragmented or even destroyed and biodiversity is reduced, then the regulatory processes no longer function well and such a virus or bacterium, a weed or even an animal can suddenly multiply very rapidly. It has fallen out of its living context and becomes a pest due to the one-sided development.

Actually, there is naturally no such thing as a pest in the plant and animal kingdom! A creature can only become a pest through circumstances—but of itself, never. It is a mistake in thinking to conclude that there are pests! Of course living beings can become harmful, but that has a lot to do with the specific situation within a particular context. Perhaps a more realistic picture is emerging.

How do we usually try tackling such pests? Of course, the whole of industrial agriculture knows only one antidote and that is eradication—in other words, war! That means inventing and producing weapons: fungicides against fungi, insecticides against insects, further pesticides against viruses, against bacteria; all kinds of things are invented and done. It is more difficult to deal with viruses. Simply eradicating them, destroying them, that is not possible. These critters can propagate so quickly that they develop resistance very easily.

To understand this problem better, let's take romaine lettuce as another example. A bad fungal disease infects romaine lettuce. It mainly occurs in monoculture. The gardeners who simply use a diversity of varieties and vegetables have few problems with this fungal disease. Already here something can be learned from this. For if I grow a single crop, I create a one-sided ecosystem and a fungal disease will come along with it, and it will be very serious. So every year a new variety has to be bred again to make it fungus resistant. It's a never-ending race! Now, in 2020, we are on the 36th breeding of lettuce. We may ask who will win this race, the fungus or the breeder? Breeding is becoming increasingly complicated. That is why seeds are becoming ever more expensive for gardeners. This is also an image of the natural phenomena we have and continue escalating.

Expansion on the Question of the Climate

Are there alternative ways of dealing with such pests or diseases? Certainly not with the usual methods: separate, isolate, try to destroy, eradicate and therewith solve the problem. I'm not saying that sometimes it doesn't have to be done this way, but it's certainly not the only and nor the best solution. However, in the long run, these methods only generate a proliferation of pests with heightened virulence. We know this about the so-called 'weeds' too; we only dampen the symptoms without getting to the root of the problem. That's why we are now going to look a little beyond Covid to climate change or, more appropriately, to 'climate breakdown': in other words our current climate situation. What kind of picture emerges?

Climate is actually an interplay of the elements: earth, water, air and warmth, and of course light plays its part too! What is happening at the moment? As a biodynamic movement, we have a worldwide network and due to the current situation, we have Zoom conferences with friends in the different countries and continents. In autumn 2020, we had a Zoom call and at the start, everyone exchanged how things were going. A partner in California said, 'I've packed my suitcase because terrible fires are raging here. At any time the fire brigade may call me and I'll have to leave the house immediately. So if I disappear from the screen, it is because I'm leaving really quickly in the car!' Then someone from northern Italy reported that there they were having terrible floods. The fields, the fruit trees, the vines, were all gone from far too much water all at once. Friends in Argentina and South Africa reported that they were living in very strict lockdowns.

So let's just consider these two aspects: happening at the same time, far too much heat and drought in one part of the world and far too much water in another. This situation, which is being experienced almost daily in the world, should really be observed and a diagnosis made. The elements themselves have become one-sidedly extreme. Each element is acting on its own, becoming a monster, a real monster! With a hurricane, for example, one speaks of the 'eye of the hurricane'. It is really like a huge creature that has completely freed itself and is now rolling over the land with incredible power and violence, destroying everything in its wake. An Indian writer describes vividly the deeply shattering experience of being in the eye of the hurricane.[2]

These phenomena are becoming progressively frequent. We don't experience it that much in Europe, but in tropical regions like the Philippines, Japan or India, it's huge. We hardly know a nice, gentle rain. No; 800 mm of rain falls all at once, for instance, as recently in the south of France at my daughter's house—not reported in the media—but nonetheless very real! Can you imagine that: 800 mm of rain in one day. Suddenly as much rain as usually falls in one year, fell in a single day. With it everything was gone: houses destroyed, fields, all gone! This is the same picture as with plants and animals, in which we destroy the coherence of life. If this becomes severe, holes appear, a number of living beings are missing, so that suddenly an organism or an element develops into a pest and spreads itself all over the place, behaving like a monster, as if it were a monster.

So what is happening? Separation is actually also occurring in the elements as well. The atmospheric interrelationships of life have been and are being disrupted. That which holds the whole world together is disintegrating; the elements are no longer working together, rather each single element or being in nature 'is simply going crazy', indicating that there are tremendous forces at work at this time.

Looking at it from yet another angle, we have to ask ourselves why we are so easily infected by this virus. The tendency has now become clearer that our organisms' immune defences are becoming weaker. Moreover, we have to extend the concept of the organism to include landscape organisms; not only all living beings, plants, animals and human, but also the landscape organisms and even the larger ecosystems, are also becoming weaker because they have been wounded, fragmented and even destroyed. These wounds, such as large bare areas in contiguous forests, are now less able to renew themselves. This is why there are two consequences: some living organisms gain the upper hand becoming 'monsters', while other organisms become increasingly weaker and even disappear.

What is our perception of nature?

Together with this second picture, we may ask: How have we looked at the world for the last five or six centuries? What have we been doing to the world to make it look as it does at present?

Interestingly, it must be said that part of this diagnosis was already made almost 100 years ago when people asked Rudolf Steiner if indications could be given for agriculture because a degeneration of the earth was then already visible or suspected due to the use of artificial fertilizers. Indeed, in 1924 the use of artificial fertilizers had already begun. Already the impression was that the plants somehow seemed to have degenerated a little and animals had become weaker. In 1922/23, a huge foot-and-mouth epidemic occurred. At that time, Rudolf Steiner, together with a veterinarian Joseph Werr (1885-1954) and Dr Eugen Kolisko (1893-1939), had already developed a remedy from coffee. Quite simply coffee—it is a great remedy! I see your joy, but of course I'm referring to it as a remedy for cows!

This diagnosis made by the farmers and agronomists formed part of the beginnings of biodynamic agriculture, particularly the degeneration of the life forces of living beings, which could already be observed at that time. In addition, since the Renaissance or since René Descartes (1596-1650), we can see that we have been viewing the world more and more as a 'machine', as a huge mega-machine: indeed the whole earth as one. All the basic elements are just chemical and physical processes and my French compatriot Descartes wrote that plants and animals are merely 'automata'. This means that when your dog barks, it's just a reaction. The dog doesn't feel anything, Descartes claimed.

Since that time, we have reduced all beings, be they the elements, plants, animals, the earth, simply everything, to mere objects, to 'things'. Everything has become a thing or is deemed a machine. One could even postulate that if we think of nature in this way, it will respond in this way! The elements, the wind, the climate are behaving like huge steam engines; they seem to be acting increasingly in a manner corresponding to the way we have thought about them. I'm not saying it is like this, but if we would just this once take this seriously as a hypothesis or an image, then as a result of our reductionist world view, we have created huge power machines and released enormous elemental forces that can no longer be controlled, just like the sorcerer's apprentice did!

Another aspect also seems very important to me. Descartes authoritatively defined as valid, that only primary qualities should be used for scientific knowledge of the world, that is, things that can be weighed, measured and counted; a view that has lasted to this day. All other qualities, the so-called secondary qualities, are illusion and do not exist objectively. This is interesting, because it means that colours, sounds, forms, scents—in other words, everything that I as a human being perceive with my senses—is considered pure illusion and not science. This is what Descartes claimed, and can be read in full detail in his book.[3]

As a consequence, what has since happened to the world? How has this view of the world—through it being repeatedly drummed into our heads—determined our world?

It has resulted in a grey, boring world scientifically. Everything consists of atoms, molecules, waves, chemical reactions or physical forces; it is uninteresting and monotonous. These are considered, however, the reality and everything else we experience directly through our senses is not reality but an illusion? At best, these are things for Sunday sermons or the poetry of poets, but is of no value whatsoever for scientists. This has indeed affected modern science and has become so extreme that at the moment, it is causing many people to wake up, to notice and question it, including many philosophers. In the meantime, a number of books have been published explaining why we need a greater sentient ecology, one that does not reduce everything to numbers and formulas,[4] and an ecology, which incorporates human beings, including their senses. This is highly interesting and quite new.

So what has happened to nature since Descartes and his definition of science? Basically, we have lost the reality of the world: the world we actually experience every day in life with our bodies and our senses, this is precisely what we have lost. We no longer believe in it. We have massively distanced ourselves from the reality of the world: the world we experience through our senses. What is the consequence of this? It is that I feel completely alone, because I have also lost part of my own nature. The consequence is that I don't have a real contact with nature, because contact with mere molecules, atoms, waves and physical forces is after all not that interesting and doesn't really work.

On the one hand, this distancing from nature has produced a great fear of it and a loneliness, and on the other hand, there is a will to master it for the benefit of humankind (hubris). Plants and animals are automata, machines, things that I as a human being can master, and even put to good use, because things can be produced from them. Fear and a desire to master taking place simultaneously gives rise to this tension. I experienced this very personally in the 70s. It was a wave back then: to be a modern farmer, you had to clear all the hedges around fields and remove all obstacles. They were nice people, but it was something like a fashion wave, automating the barn and conducting mass livestock farming; a kind of war-like agriculture arose out of this duality of man and nature, out of this distancing. Every one of us can also look for this sensation within themselves: when have I, in myself, experienced a feeling of separation from nature and my environment?

Professionally, I do lots of botanical guided tours and courses with students and course participants. When I point out a plant, what questions do they always ask first? Is it good? Is it a weed? What is it useful for? And today as well, can you eat it? For today, it has become fashionable to want to eat wild plants. However, my main message to them is—before you eat it, perhaps you should greet it and ask, who are you? So before anything else really get to know it because the plant is a living being. We are all looking for contact and knowledge and don't just want to eat everything or make it useful.

This separation of man and nature has not always existed. It is said that the forest of Amazonia has a huge natural diversity of its own. As recent research has increasingly shown this was actually a cultivated forest.[5] The people who lived there very sensitively cultivated and shaped the whole forest, nurtured plants, enriched the soil and developed all kinds of elaborate designs. They were very closely connected to their natural environment and actually cultivated the diversity. It is the same in Australia, where some people thought the 'poor' aborigines were stuck in the Stone Age. Not true! A new book has been published that shows how the Aborigines also developed a particular kind of agriculture, by an active creation and cultivation of nature in a very special way.[6] This was based on an intimate relationship with nature, however, and not at all on a passive attitude in the sense of 'I just gather and hunt in nature and do nothing myself'.

Our modern view is quite obviously an ignorantly constructed one in which humans and nature are separated and that it would be best for both to remain so. This has consequences for nature conservation, for instance, as the view still persists that the best for nature conservation is to do nothing. This was never the case in reality and history shows us that human beings can do positive work within nature.

The Change of Attitude in Biodynamic Agriculture

This is also the impulse of biodynamic agriculture, especially with regard to the aim of transforming our attitude towards nature. During conversion courses, farmers often say that the numerous biodynamic measures with the spraying of preparations, the creation of compost, among others, are not easy, but can be learned quite well. However, the requirement to view nature differently, to develop a different approach to nature really takes time. It is not easy at all, because it is an inner process.

We need to observe and notice our own thinking again and again, for instance, when it comes to doing something about fungal pests or weeds that are to be cleared. For how can a different attitude be developed to deal with nature in such a way that we develop healthy forces and create conducive localities? The most important principle is simply that we honour nature beings and reject that they are things or machine automatons. Plants and animals should be seen as organisms and beings, as they are: as their 'inner' nature is. This is what nature-based agriculture actually means.

What is soil, for example? That one really looks without preconceived ideas and discovers what this particular piece of ground is, what its plants are, what its animals are [is important]. That you ask this question every time, for example with wheat, 'Who are you?' Or, an apple tree, 'Who are you?'. That's what we farmers do with our cows, 'Who are you?' Or, every dog owner really asks their dog, 'Who are you?' With this attitude, we are able to open ourselves up to the perception of the other being. Goethean science is extremely helpful here: open yourself up and ask, Who are you? Then develop the criteria necessary for cooperation through this process of getting to know each other. I think that would really be the first principle—because in agriculture, we mainly work with living beings.

In this way, we can then also perceive the earth as a whole, as a living being—instead of the ingrained view of nature as a kind of machine.

In addition to the concept of the earth as a whole, we also have to try working on an inner attitude towards the living organism. What this means is interesting. Rudolf Steiner begins his agricultural course with this concept and emphasizes that all agriculture should be or become a self-contained organism. For a long time, it was thought that this was a unique view of biodynamics. Today, however, a number of researchers have noticed how this actually makes sense, because there are multiple organisms living around us, each one made up of a diversity of living beings.

I'll just take the beehive as an example, a bee community. This is a wonderful organism in which many beings work together. The single bee alone is perhaps not that intelligent. But all together, the bee is extraordinarily intelligent and wise. This is really an impressive picture and their working relationships are invisible. This can also be used as a very good image for the agricultural organism, for example. Today we realize that each one of us is just such an organism. I contain more microorganisms, more bacteria and viruses than the number of my own cells. So I am also such a complete whole organism in which a great variety of beings are active. Therefore the idea is not so crazy when speaking of the agricultural organism; a biodynamic farm is made up of just such a diversity of beings.

Nevertheless, the question then arises as to how to create a kind of boundary to the surrounding areas. In working with the living, we work with the principle of semi-permeable membranes. These are semi-open sheaths just as the skin is, for example, which clearly creates a boundary. Something similar also applies to our breath. So how do we create such semi-permeable membranes for an agricultural organism at various levels? This can be explored and physically defined or designed with hedges, with trees or a stream. But it has to be created!

This can also be done with the animals by rotating the animals between the different meadows and different areas so that they 'enliven' the landscape, thereby creating a soul sheath for a farm. One immediately feels that animals are at home here and not somewhere else.

Such a sheath can also be created spiritually in that the people who are responsible there really raise their consciousness again and again: this is the organism that I care for, this is my responsibility. This creates unity and really yields great strength.

An interesting experience among the gardeners working on the large Goetheanum grounds is the following. When their consciousness is spread all over the whole property, nothing happens. However, if for instance, they have much to do in the summertime or at other high peak times of their work, some areas slip out of their consciousness, then something can happen. So these kinds of approaches also form a kind of sheath, through intention and awareness of the whole organism. Of course, these words are being spoken quickly, but it takes a lot of step by step work to manifest and form it through the course of time. How this works in detail can now be seen on very old biodynamically managed farms, such as the Marienhöhe farm near Berlin—how such an agricultural organism has been maintained and developed over 90 years through all its ups and downs.[7]

This farm is really a living being where the farmers have actively conversed about the design process over years. This is a completely different way of thinking, than the approach in industrial production.

How do I design an organism in practice?

An organism simply needs organs, as is the case with us humans, for the head and the feet are very different. In French we say, 'I think like a foot', which means very badly and you'd better not do that! How does one create and then care for such organs? This brings us right back to the topic we had at the beginning—you don't detach, but bring and connect the different elements by creating places, for example, where good humus-rich earth, water, shade and a mild warmth come together, like a brook or riverside pasture.

Then there should be another place on the farm organism where warmth, light and dryness can intentionally come together or develop, of course especially where the preconditions already exist. This is how you create strong polarities in the farm organism, places where the different elements meet and interpenetrate. Rudolf Steiner is very clear about this. It is much better to consciously preserve and use 5 or 10 % of the land for such natural ecosystems than to try cultivating every area of the farm. Some organisms will thrive better in a humid area such as a wetland, others in a forest, others in dry pastures or in a dry wall. In each of these natural elements, preserved and maintained or newly designed, an individualized habitat is created for many different animals and plants to settle and live in. Then Steiner adds something illuminating—for the bacteria, microorganisms, fungi and all the pests that are found in agriculture, you have to find a place for them on the farm, and that place, for example, is a wetland with multiple fungi and microorganisms. If they have a place there, you will then have fewer problems in the fields.

When I say this to the farmers in my conversion courses, the reaction is, 'You're crazy! We've had so much trouble getting rid of mushrooms. Every year we burn them and yet everything still gets ruined, and now you want us to establish them, take care of their habitat? So please, just not that!'

It goes even further. Some time ago I was on a farm in Normandy, in a humid area of western France. There is an agricultural organism that has been shaped and formed for 40 years with a lot of forest, humid areas, a stream and a lot of vegetable cultivation, as well as cows and a pig. When I asked the farmer, 'How is it? Do you have fungal diseases or pest problems with your vegetables?' His answer was, 'No, a little yes, but normal and it's not much!' However, his neighbouring farmers have intense problems, even neighbours who work organically-biologically have more problems with insects than he does. They need to use increasing numbers of insect nets and everyone complains about it every time they meet. But this farming couple has few, loses little yield, and does not experience problematic fungi and insects. Why is that? If you observe life on the farm and all the goings-on, you soon notice a sense of joy. Many Parisians come here; it's a sought-after place for them to buy their fresh, biodynamic vegetables. In addition, there are their children playing around the place. Many people make something of a pilgrimage to this place, enjoying the animals, the farm and their children. There is a sense of real enjoyment there at all times, which strongly radiates from this farming family.

Why do I say this? Because the whole mood of this organism simply radiates joy. This makes a great deal of difference! I really think that the joy, this happy and positive mood has an effect. This farm organism has a direct soulful effect on its environment—and this has more of a positive impact than constantly dwelling on what could be done against the insects, when action should be taken against the fungi, how much these pests will cost me, and so on. This does not bring any joy at all! At this farm, however, joy is present and a positive atmosphere is created.

Another experience: on this farm there is a pig living right next to the vegetables in a marshy area. It gets all the non-usable leftovers from growing vegetables. The existing fence had had no electricity for six months and was by then hardly visible, lying in the grass on the ground. The pig had lived there for a whole six months without a fence and never left, for example, by going to the vegetable fields! It got food all the time and indeed, the pig was happy and had fun. It even used hay for making its own mattress. It cuts grass with its teeth, lets it dry and that is now its mattress. Unbelievable!

With this I want to point out that animals are of course a very important part of such an agricultural organism. Animals belong to every ecosystem, to every plant, and the integration of animals is of course a great challenge for modern farmers, because of the dominance of monoculture.

I know the situation well because I am often with winegrowers. Many winegrowers now want to convert to biodynamics, in France, Spain, Italy, Switzerland and Germany—in fact, all over the world! It's really terrible for them, because they are pure monoculturists, with very unnatural ecosystems where the vines are constantly in competition with each other. They all need the same nutrients from the soil at the same time. This is a full-scale agricultural industry. Then the biodynamicists come along and say, 'That's not possible, you have to create an agricultural organism!' How would you then manage to integrate animals? This requires a huge rethink for winegrowers: that animals are welcome and not harmful. Winegrowers fight fungi and other diseases and all kinds of pests every single day. It is one huge disaster being a winegrower! From March to July, he always has to think: What should I be spraying now so that the grapes continue growing? It is an enormous challenge.

So he has to ask himself, 'How can I learn to think positively now? I have night butterflies, for example. Good, then I have to make sure that I let bats back into the vineyards. The bats eat night butterflies, then I won't have to worry about what to do about that. But to attract bats, I need hedges, because bats don't fly around in completely open landscapes. I therefore have to start designing my landscape.' Then the next question arises, 'How can I integrate animals?' Winegrowers then mostly try to work in partnerships, that is with colleagues who have animals: with a beekeeper for keeping bees, with a shepherd for integrating sheep, and much more. What do winegrowers then say after this process? 'Yes, the whole viticulture is becoming healthier, step by step. Not immediately of course, because it takes time.'

Sometimes, very surprisingly, problems are solved by themselves. For instance, in orchards, there are often a lot of field mice. When I introduce sheep into my orchard, the field mice disappear. Why? Because the sheep's footprints on the surface of the soil are unpleasant for field mice. They then prefer to leave on their own. Yes, that is a sensible natural solution, but we didn't know that before. We discover all kinds of things like this all the time in our dealings with the agricultural organism and nature beings.

What do these examples mean on a large scale? As modern human beings, we are increasingly shaping all these networks and are actively responsible for these cycles. It was different in the past. They no longer come about so easily on their own anymore. Through these methods, we promote biodiversity, which we integrate by basically developing an attitude towards every living being and say, 'Welcome to the farm! You will find your place on our farm and you will be given a task appropriate to *you*.' In this way very many problems are solved, step by step. Not all of them, of course, because there are always others. Very important to understand though is that problems no longer create fear.

However, how should this overall evolution continue and what is its goal? These are big questions. I personally believe in going in the direction of individualization. If you closely follow the unfolding situation, animals and plants are becoming increasingly individualized. This is also a principle that we have in biodynamic agriculture for the agricultural organism. We even pursue the goal of achieving individuality on the farm. This means that we strive for each place to develop its very own quality, a kind of genius loci that can express itself in the natural events in the course of the year, a genius loci that is very unique to every farm, but which is to a certain extent also invisible. An individual life context that may and should express itself and which becomes more and more differentiated and developed through time. There are of course many various ways of doing this, but biodynamic agriculture is one that has been tried and tested for almost 100 years.

This is partly due to the spraying of our preparations, two biodynamic preparations that we use in very small quantities. One is made from 'horn manure' for the soil, which is sprayed in the evening, and the other is 'horn silica', very finely diluted, which is sprayed on the leaves and fruit in the early morning. If you do this regularly—and here I come back to the winegrowers, because the vine is a very sensitive plant and reacts sensitively to these preparations—the winegrowers say that they can directly experience its positive effects. Either in the posture of the vine and in the shine of the leaves, but even more strongly in the product, the wine itself. The wine simply has a stronger 'terroir taste'; it is simply more specific. This observation is confirmed, for example, by the best winemaker in the world in 2000, Olivier Poussier.[8] For the wine, this specific individuality is more interesting than a standardized quality. Of course, this can also be found and tasted in a carrot, but for this you have to taste the carrots as finely as the wine. One can also taste the positive effect very well with medicinal plants.

When we spray these preparations, the identity of the plant is strengthened through an intensified relationship with the particular place, and in its nutritive or healing power. For medicinal plants, the possibility of creating a better remedy is developed. We don't spray directly on the plant. We spray a little all over the ground. That means we actually spray atmospheric preparations that strengthen and harmonize the connections, the relationships, creating a coherence with the environment. The relationship between soil and roots, for example, improves. The roots can develop better under the soil and the leaves, flowers and fruits above the soil through the 'horn silica' preparation. This improves the relationship to the light. The plant can process the light better and gets a fine differentiated quality of the light. In other words—we create healthy atmospheres.

Then we have another biodynamic preparation, the valerian preparation. If it has hailed, bringing an influence that is too cold, or if there was frost, the very fine flower juice of valerian is used, very finely diluted and also sprayed. It smells so wonderful and it really helps against the effect of hail on the plants, or, if there was a night frost, then the plants are able to regenerate more easily after the frost. We thereby create a warm atmosphere; that is the idea of this process. A farmer recently told me that he had the impression that he was perfuming his fields with this preparation. A fragrance creates an atmosphere. Sometimes in a room you say, 'it smells good here!' But it doesn't always have to be physical, it can also be spiritual. So in the atmosphere, one also creates soul moods that can be either positive or negative.

This is a very big aspect for me for the future. Because our problems, for instance, COVID-19, pollution, climate problems are all atmospheric problems as well. They come from the periphery and are present everywhere. Many of these issues are very real breathing problems that affect not only the physiology but also the psychology and even the spiritual: I become anxious, get stressed and so on. It is interesting how the phrase 'I can't breathe' has become a signature of our times. I am no longer connected to my environment, to nature, to society. The problem comes from the atmosphere. Covid, where is it? I don't know. Air pollution, where is it? I don't know. And so on.

I am convinced that our next step is to further develop our behaviour, through thinking in an 'organism' way, an integration of nature and ourselves, bearing a message of welcome to nature beings, by according them a worthy place. For this we need to think and feel much more atmospherically! This means learning to behold the invisible. What is the 'atmospheric' between things? What envelops us and how? So we gradually understand better how to sense the atmosphere—not in a focused, nor causal, but in a holistic way. This is a very big challenge for the future!

In a meeting of trainers, a colleague gave input and stated, 'Yes, science shows that biodynamics works, that the preparations work.' He had presented a series of scientific analyses. The other participants replied, 'But you know, we don't need that for our farmers. They are not that interested in the scientific proof.' Why do they want to convert to biodynamic? Because there is a good atmosphere on the farms and because there is a good atmosphere at biodynamic meetings. This is no joke! Today many farmers very often become resigned. A calamity indeed! The current situation in agriculture is really grim. Many farmers are deeply suffering.

How can we help all farmers to really enjoy their work again? This is very important. Indeed, almost all young people studying agriculture today want to work organically or biodynamically, and almost none conventionally. So there is a lot of hope, and a certain change of world view is already underway among many young people. They need our help and support, of course, by buying their products and supporting them in various ways.

It was important for me to generate a little enthusiasm and delight towards a fundamental change in agriculture and in our outlook to nature.

Slightly edited lecture from 26 October 2020 in the Goetheanum carpentry workshop.

[1] Morand Serge: L'homme, la faune sauvage et la peste. *Fayard. 2020.* (French only.)
[2] Amitav Ghosh: The Great Derangement. Colonialism, Climate Disaster and Human Delusion. University of Chicago Press, September 14, 2016.
[3] René Descartes: *Meditations and Other Metaphysical Writings.* Penguin Classics, 1999.
[4] Jacques Tassin: *Pour une écologie du sensible…* Published by Odile Jacob, 2020 in French. A summary 'For an Ecology of the Sentient. Weaving a new connection with nature' in English is available at:
https://en.odilejacob.fr/catalogue/science/environment-sustainable-development/for-an-ecology-of-the-sentient_9782738148964.php

5. *Hundreds of years later, plants domesticated by ancient civilizations still dominate in the Amazon*. By Erik Stokstad, 2, Mar. 2017 on *Science Mag* (AAAS) https://www.sciencemag.org/news/2017/03/hundreds-years-later-plants-domesticated-ancient-civilizations-still-dominate-amazon
6. Bruce Pascoe, *Dark Emu: Aboriginal Australia and the Birth of Agriculture*. Magabala Books. Australia, 2018.
7. Fridtjof Albert: Marienhöhe—Three generations of work on building up soil fertility. English *Report of the Agriculture Conference* at the Goetheanum in Dornach, Switzerland, 2017. https://www.sektion-landwirtschaft.org/fileadmin/SLW/Literatur/Tagungsdoku/2017/Landwirte-Tagung-2017-EN.pdf
8. Hors-série N° 36, Revue du Vin de France.

Peter Selg

'Building a Bridge to Right-Wing Extremism?' On Anthroposophy in the Time of National Socialism

Dear Ladies and Gentlemen,
9th November is a date that affects us all very much, especially those of us who come from Germany. 9th November has been inscribed in German history in multiple ways, in the light and in the shadows [1], but 82 years ago today was a dark day in German history, one that marked the beginning of one of the greatest catastrophes of all times. 9 November 1938 was the day of the Reich's Chrystal Night, which the national socialists famously called the 'Reichskristallnacht'. What happened that night did not have anything to do with crystals, even though an infinite number of things were broken, and not only panes of glass. It was the day and night that synagogues were destroyed and Jewish shops were looted, more than 7,000 businesses and countless private homes; about 26,000 Jews were deported to the concentration camps that already existed by then: Dachau, Buchenwald, Sachsenhausen. A date that is commemorated today in Germany and in the world—a relevant date, indeed, a highly relevant date!

It is also the 100th year of Paul Celan's birth. He travelled by train from Krakow to Berlin via Auschwitz, Oswieçim, on 9 November 1938. Later he wrote these lines of poetry:

Upon arrival in Berlin,
via Krakow,
you were met at the station
by a plume of smoke,
tomorrow's smoke already.[2]

The Reich Pogrom Night was the beginning of the many catastrophic things that came from 'tomorrow' and yet had been in preparation for years, had been systematically prepared. It is good to talk about right-wing extremism on an evening like this, about right-wing extremism and racism. In this context, you may ask, what does anthroposophy have to do with this? In 2019, on the centenary of Waldorf schools (the first school was founded and opened in Stuttgart in September 1919), various German newspapers, television and radio broadcasting services, as well social media, reported that the schools had, for the most part, done respectable educational work; one flaw, however, was the well-known nationalism and anti-Semitism of their founder Rudolf Steiner. Almost exactly one year later, during the course of street protests against the German government's Corona measures, especially after the large demonstration in Berlin at the end of August 2020, similar but much harsher press reports appeared. For example, on 1 September 2020 in the *Zeit Online* and shortly afterwards in the left-wing magazine, *Konkret*, according to which, a part of the Waldorf 'scene' had taken part in the protests infiltrated by the right-wing extremists, presumably due to its long-standing affinity of anthroposophical circles to right-wing extremist ideas; *Zeit Online* said that anthroposophists were 'building bridges' to right-wing extremism on 1 September 2020, the anniversary of the beginning of the Second World War.

In view of such far-reaching and, in my opinion, grossly inaccurate assertions—such as, anthroposophists 'building bridges' to right-wing extremism—it seemed meaningful to us, the Goetheanum Leadership, to include the topic in this evening's lecture, on this particular day, in the series on the *Signature of Our Time*. We will address the question: What is the state of historical research on the topic of anthroposophy and national socialism? We will also address the question: What are the challenges we face today, our contemporary challenges? This is far too comprehensive a topic for a one-hour lecture, truly too big an undertaking. This is also why I have decided to speak to you very personally, as Paul Celan once said, 'out of the inclination of one's own existence'. He did not mean 'inclination' in the sense of preference or affinity, but in the sense of one's own view having become so, having evolved this way and not another. We all have a history, a historical existence, are ourselves this historical existence, a distinctly shaped existence written in space and through time.

Personal emphases and the staggering allegations

I was born in Germany in 1963, 18 years after the end of the world war, German fascism and the Holocaust. I grew up in this time and in this country, in a period that was still characterized by displacement and a slow process of coming to terms with the past. Since then, I have dealt with the topic of this evening's talk a lot, but I do not claim to have an objectively valid, or, an 'absolute' verdict to lay before you. Understand my contribution much more as a personal search and a sifting through, but also as a personal inner inquiry. After all, we all continually and very differently emphasize things that are important to us and which may not be such to another person. I suppose, we have to do that and it seems legitimate to me. One should acknowledge one's own emphases and priorities; they are not objective history, but are that which one is inclined to see particularly clearly from one's own unique perspective, out of one's own existence, what is important and catches one's eye, things to which one ascribes significance, crucial significance, and probably even, decisive significance.

However, I do not at all mean that it is *only* about personal accentuation. At the same time, I insist on the presence of historical knowledge, historical knowledge, which is essential. Thus, I also wrote about this to the editors of *Zeit Online* to correct their article of 1 September 2020, but have received no reply to date. Yes, there is objective historical knowledge about Rudolf Steiner and the anthroposophical movement and national socialism—and they should have taken this knowledge into account and internalized it before writing such an article on such a historically significant and heavily laden day as 1 September, and especially before putting it out into the world. It is, as I have said, legitimate to have one's own perspective; we should grant that right to all authors, even if we do not like and are not comfortable with theirs. That is what I am going to do tonight, taking the liberty to do. However, this emphasizing, highlighting, accentuating needs to be done within an overall context, that one has first to take cognisance of, know and understand—even if it is not easy or quick to read and penetrate, and even if it makes hermeneutic demands.

My lecture will have three parts, three sections. First, I would like to talk about Rudolf Steiner, about his attitude, his statements, his position on these pressing questions of nationalism and right-wing extremism, racism and anti-Semitism, as far as this is possible in the short time available. Second, I will then move on to the question of how anthroposophists—the people who attached significance and a positive meaning to Rudolf Steiner's life work—behaved from 1933, eight years after his death on 30 March 1925, to the national socialist's takeover of power, and what they then did in the following twelve years until 1945. Finally, in a third and last step, I would like to outline where I see some of the tasks, challenges and difficulties associated with the subject of anthroposophy and national socialism, what documentary work and reflection still needs to be achieved.

There is really a lot to consider in this whole discussion, and my sketch will be insufficient and provisional, but I would like to tackle it nonetheless. Here in Switzerland, the topic of this evening is somewhat further away and does not concern people so existentially. And yet—what does concern us in Germany, we owe to a person like Micha Brumlik, who was born in Davos in 1947 and went to Germany as a child, in 1952. He was later a professor of education in Frankfurt am Main and presented ground-breaking studies on anti-Semitism, held the Franz Rosenzweig Visiting Professorship in Kassel and received the Buber-Rosenzweig Medal. He came from Switzerland and yet did this valuable work in Germany, this enormous reappraisal of anti-Semitism, which was by no means limited to Germany. According to Primo Levi, anti-Semitism had no greater prevalence in Germany in 1933 than in any other European country. However in Germany, during the Nazi era, it led from systematic registration of the Jewish population, to persecution and genocide, a killing of unimaginable proportions, a dynamic of extermination that world history had never known or seen before in this way.

I will try to unfold my topic—anthroposophy and national socialism or the supposed 'bridging to right-wing extremism'—to you this evening in a calm way, despite the mere one-hour allotted speaking time. However, I would like to concede from the outset that this is more than difficult and will not and cannot be fully achieved. The whole topic is the opposite of calm; it is deeply disturbing—especially in view of the growing right-wing nationalistic, authoritarian, regressive, sometimes even totalitarian forces at work worldwide. After all, we have been observing a highly alarming shift to the right in societies again for many years, and it has been closely examined in many publications. A party like the AfD has become a political force in Germany, a factor to be taken seriously—just imagine! And things are no better in other countries, on the contrary.

However, ultimately, the subject is highly disturbing 'in its own right' because, with regard to anthroposophy, the assertions being made, such as those mentioned above (*Zeit Online*), are accusations that take the breath away of those with knowledge. Last year a book was published in Germany with the title, *Anthroposophy: A Short Critique*. Its author, André Sebastiani, claims that anthroposophy is 'elitist, dogmatic, irrational, esoteric, racist and an anti-enlightenment worldview'. Rudolf Steiner was a 'radical anti-Semite' and anthroposophy is, at its core and in all its details, an 'inhuman ideology'[3]. For someone who, like me, has studied Rudolf Steiner's complete works for decades, this is a shocking conclusion to come to. I do not mean with regard to Rudolf Steiner, but with regard to the mentioned publication. How can someone assert and spread such things, how can they come to such judgements and publish them in a style and diction as if someone with knowledge? Has the author ever studied anthroposophy, ever taken cognisance of anthroposophical anthropology with its deep humanity and social spirit? It is distressing piece, and extensively available in bookshops. It is also shocking when Helmut Zander claims that Rudolf Steiner belongs to the 'intellectual background and superstructure of the German tragedy'[4], implying that he, Steiner, was one of the preparers, one of the forerunners of German national socialism, even though he was no longer alive in 1933. Like Hitler, he had come from Austria, with comparable family backgrounds, from German nationalist and anti-clerical backgrounds; together they set off, so to speak, as can be read in Helmut Zander's widespread, optimally positioned, extremely popular biography of Steiner (Steiner and Hitler, both 'Austrian, German national, anticlerical'[5]). Shocking, shockingly false and suggestive, even grotesque in this combination and simplification, is what I say. What kind of sentiment and prejudice is being evoked to a public that believes such 'expert' opinions? In the USA, Professor Staudenmaier writes that anthroposophists benefited massively from the Nazi regime up to a certain point. Anthroposophical doctors would have been able to make a 'career' from 1933 to 1945, would have enjoyed certain privileges, even if in the end they lost out ...[6]

In view of what we know historically about the relationships of anthroposophy and anthroposophists—including individuals from within anthroposophical specialist fields such as doctors—to the Nazi regime 1933-1945 [7], these are thoroughly audacious accusations and a calm lecture dealing with them is no easy undertaking. Particularly as they are accusations with huge media impact, which are seized upon by journalists and taken at face value by the public, as indeed was the case in the *Zeit Online's* article on 1 September, by an author who has certainly never examined the complete works of Rudolf Steiner and assessed its content. These claims continue to have an effect in the 'echo chambers of the internet', are reinforced and spread through mass media, and which we, in the Anthroposophical Society and movement, have almost no chance of dispelling, despite our specific historical research and knowledge.

Dear friends, it is truly not meant to be an apologetic evening, but I also cannot ignore the fact that such distortions abound, that the terrain we are treading is not an open and free one. When we stand up for anthroposophy, we have to deal with such defamation, and this diminishes the calm composure needed for self-critical historical reflection—as if there could ever be thought of 'composure' when dealing with national socialism and the Holocaust, even independent of anthroposophy and the accusations mentioned.

I said, it needs calm composure for self-critical historical contemplation. For this would ultimately be the goal: to work out how anthroposophists behaved from 1933 to 1945, including how they adapted, became followers, were driven by fear, where they were in denial of themselves and anthroposophy, or where they were wanting and had fallen short of themselves. They were not all heroes, martyrs, resistance and spiritual fighters, determined to the end! And the anthroposophical institutions were under great pressure, wanted to continue working, but were always on the verge of being prohibited and close to the abyss. After 1945, like all other social groupings, they spoke little of this, because it was apparently now over and because it had been in part shameful—because this terrible time had had a traumatizing effect, but also because the differences of opinion and behaviour towards the regime had left deep cracks in society and in its communities. All these things need to be critically examined and critically discussed—and this especially with regard to the present and the future.

How do we, as anthroposophists, behave under fear and pressure, threat and persecution in the world; how much civil courage does each one of us have; what kind of non-conformist behaviour are we capable of in civilization and the social crises of our time, from ecology to threats to democracy, from refugees to climate protection? What, if anything, do we risk by advocating a new and radically different economic system and behaviour that breaks with the dogma of unconditional economic growth and its externalized consequential damage? Critical examination of the behaviour of our predecessors under the aegis of a highly authoritarian system could be of educational value for us.

However, how can this debate be conducted calmly and prudently when it is publicly claimed from the outset that anthroposophy itself and Steiner himself are immanent to 'radical anti-Semitism', German nationalism and authoritarianism, with an ideological dogmatic 'contempt for humanity'? If one has to defend oneself against such accusations—which implicitly assume that people who still associate with anthroposophy today have a natural proximity to right-wing radicalism—then there is a danger that self-criticism will wither away, be put on a back burner, come up short or perhaps eventually not at all. This is how contradictions increase, social tension and polarization rises, 'black' and 'white' continue bitterly opposed to each other. This is a problem, also a methodological one.

Rudolf Steiner's Approach

Without being able to resolve the methodological problem in any way this evening, but in full consciousness of it, I come to the first part of my talk. I would like to underline in short, as highlighted earlier, and at least hint at a few of Rudolf Steiner's core approaches to this whole complex theme, beginning with the *Philosophy of Freedom*, his main philosophical work. In this book—well-worth reading—which he considered valid throughout his life and which was continually reprinted until 1925, you will find his commitment to individualism; it was the decisive principle for Rudolf Steiner from beginning to end. The individual is capable of cognition. It is capable of insight and, on the basis of its insights is also able to advance to an individual morality, is able to attain an individual cognition of truth in a situation, a situational recognition of truth, making this the determining factor for its own actions. The individual is able to detach itself more and more from external determination, including from the external collective, and to arrive at self-determination. Rudolf Steiner saw the future development of humanity in this. He wrote in his *Philosophy of Freedom*[8]:

> Determining the individual according to the laws of his genus ceases where the sphere of freedom (in thinking and acting) begins.

There is something that imprints upon us, that preconditions us through the situation into which we are born, in our physical bodies, social and cultural situations; with our unequal preconditioning, we socialize and grow up unequally. From an anthropological perspective, this concerns, by far, not only the body in the narrower sense, but more broadly, our habits and sentiments, our opinions and prejudices. Yet, our real development lies in detaching from what has been before and from what has been inherited, in overcoming and in becoming oneself, albeit not in a solipsistic and egomaniacal way, but rather through our experiences in relationships, with 'you'. In Steiner's view, this is the true worth of every human being in his or her incarnation. This was one of his central points.

These external determinations—these imprints by which history has long been determined, including the 'folk' and the 'races', which were still big issues at that time, in Steiner's lifetime—must disappear and be overcome, Rudolf Steiner emphasized again and again. Their significance is decreasing, had indeed already ceased at the beginning of the twentieth century, having played an ever diminishing role. The *worst* was the attempt to continue with these old concepts, to *conserve* them, to keep them alive artificially, or to derive or postulate things like racial ideologies from them. Steiner called these 'raw monstrosities' and deeply 'materialistic'.[9]

Why materialistic, you may ask? Because it is precisely in the sense of this way of thinking, which claims that the material body or the physical body, the *genes*, the neurobiology, the *race* and the *blood* into which we are born, shapes and ultimately defines us, which *determines* the human being—this is materialistic. Yes, we arrive on earth in different life situations, but we transform these inborn characteristics by individualizing them more and more, including our genetics in an 'epigenetic' way. This is our task, Steiner emphasized.

Dear friends, I think it was significant that someone in the first decade of the twentieth century said that the chasms between nations would and must disappear more and more, and that human cooperation that goes beyond nations—which should as such sink into complete *insignificance*—would begin. It is significant when someone like Steiner claims in the first quarter of the twentieth century that the destiny of the whole of humanity can no longer be determined, nor solved nationally, he literally means, '... since in the near future the destiny of humanity will, to a much greater degree than has hitherto been the case, bring people together in a common human mission'.[10] Now today, one may be allergic to the concept of 'mission' or 'human mission', and perhaps one should be—the missionary has mainly become a nightmare for us. However, in this context Rudolf Steiner simply means: tasks, human tasks, which are to be addressed, mastered and solved.

Human challenges and tasks, such as those we have moved through in this series of lectures on the *Signature of Our Time*, from ecology and climate change—'climate change' is too gentle a word for the catastrophe that is becoming ever more apparent—to other enormous problems that no country can escape from, to the issue of 'pandemics'. We can no longer escape from the climate problem, perhaps from the Paris Agreement, but we cannot escape the problem, nor from the many other problems either: the problems of the zoonoses of our day, the consequential damage from the loss of biodiversity, the collapse of our ecosystems, and much more. This was already clear to Rudolf Steiner in the first decades of the twentieth century, even if not in all its details. He was not the only one who saw it this way either, but he saw it very, very clearly—and it is worthwhile engaging with his clarity in terms of content.

People matter—not nations or skin colour

To return to the main theme. That which is encompassed in words such as genus, race or nation had, according to Steiner, a sustaining significance in the past, but has become much less important. In the future, the human being alone will matter, not as the wearer of a skin colour or a confession or nation, but the human being *as human being*. In a summary, Wenzel Michael Götte once wrote:

> A common thread runs through Rudolf Steiner's written and oral work. It says, learn to recognize the human being in the other. The human being of spiritual origin. For to find human beings means to connect with them spiritually. The individuality develops by surrendering to the universal human being. That is the goal. In as much as the individuality cannot exist on earth without flesh, but is 'incarnated' in inverted commas, it experiences resistance and support, talents and handicaps are also inherited, it stands in space, in a particular place, as it lives life through time, in a particular era. The diversity of innate characteristics are a reality that cannot be denied. For they are definitely connected to an individual's destiny. This has to be understood. A deeper spiritual understanding, however, leads to the thought that Rudolf Steiner formulated in this way: 'You are human, with all the other people of the earth'. This is the spiritual breath that flows through Rudolf Steiner's work.[11]

I would like to remark in parenthesis, that one can well understand why Christian Morgenstern wrote a letter to Oslo in 1912 proposing that Rudolf Steiner receive the Nobel Peace Prize. What would Christian Morgenstern have said—who, unlike authors like Sebastiani really knew Steiner's work very well—what would he have said to the assertion that anthroposophy is an 'elitist, dogmatic, irrational, esoteric, racist, anti-Enlightenment worldview'? I don't want to know and I am pleased that he, Christian Morgenstern, this noble and fine spirit, was spared such 'literature'.

So far I have only highlighted and underscored a few points. However, I would like to see such statements as the ones I have mentioned being discussed in public, even by non-anthroposophists, and associated with the name,

Rudolf Steiner, and not hushed up.[12] They were astonishing for his period, for the first quarter of the twentieth century. When someone comes along in the middle of the First World War, in 1917, when everything was so focused on nations, tremendously focused, and formulates a viewpoint for a 'social threefolding', this is more than unusual; it is surprising or 'extraordinary', in Goethe's sense, 'worthy of note'. The concept of 'social threefolding' implies, as I am sure you all know, the overcoming of the nation state, the overcoming of centralized state systems, of regulating from the 'top-down'—in favour of a horizontal federalism with autonomous subdivisions. Education, spiritual life, cultural life, religion, these are extremely relevant aspects, indeed, they are essentials of being human, which cannot simply be determined by the economy or by the political state system. Rudolf Steiner formulated—in 1917!—in the midst of the world war, the prospect of a complete reorganization of society—in the place of enforced nation states and power complexes—which included new, associative forms of economy. He outlined this in a structured proposal for a peace settlement. For at that time, the question was very much how to end the war and the deadlocked situation, how to solve it. At the time, Steiner was one of the first to make crystal-clear how explosive and problematic Woodrow Wilson's policy of nation states was, which sounded so good on the surface ('the right of self-determination of the people'). Today, this stands clearly before the eyes of historians, but back then? Wilson's racism is also an issue today, universities bearing his name are being renamed; on the other hand, at the time, he was regarded as a 'shining human light', was celebrated in an almost messianic way in Europe and was awarded the Nobel Peace Prize. Rudolf Steiner took a clear counter-position to him, as can be read in detail. He warned again and again—to the astonishment of his listeners at the time—against a further intensification of the nationalist and ethnic thinking of Wilson's initiatives.[13] Today we know what misery Wilson's nation-state policy caused, especially in Eastern Europe—but there is no mention anywhere of Steiner's warnings against this. Nor anywhere else! Here in this room, in the carpenter's workshop, he said on 26 October 1917:

> Someone who speaks of the ideals of race and nation and of tribal membership today is speaking of impulses which are part of the decline of humanity.[14]

According to Steiner, this is precisely what will lead to the abyss. Three years later, again in this place, he said:

> However, as many more nation states are set up, so many more seeds of destruction are provided.[15]

Yes, they are individual sentences, I know, but one can informally explain what Steiner's concept of society was; a society to be reorganized, against the background of which he said something like this. This is not an isolated quotation, but a 'text', as I said in the introduction, which intentionally culminates in such formulations. In September 1924, in one of his last lectures here in the carpentry workshop (*Schreinerei*), Rudolf Steiner emphasized:

> The way that men are striving towards races and nations, and the way that they basically want to bury all cosmopolitanism today is really quite terrible.[16]

There were not so many influential thinkers in Germany in 1924, who spoke of the 'burial' of cosmopolitan thinking and saw it as a deep tragedy.

Working Against Anti-Semitism

And because he saw it that way, Steiner's position towards the anti-Semitism of his time—and anti-Semitism in general—was also extraordinarily clear and explicit. Are the authors, who attest to Steiner's 'radical anti-Semitism' and right-wing nationalist thinking and who publicly spread their allegations, aware that he published nine essays in Berlin journals against anti-Semitism as early as the beginning of the twentieth century, from 1900 to 1902?[17] Already at the age of 19, in 1880, he called anti-Semitism (using the example of the philosopher Eugen Dühring) 'barbaric nonsense' and a 'mockery' of all the 'educational achievements of modern times', and spoke of the 'inferiority of the spirit' and 'lack of ethical judgement'.[18] What Rudolf Steiner saw in the 1890s—and later—in a decidedly critical light was Zionism and the intention of founding a Jewish state in Israel. He witnessed the beginning of the Zionist movement, and he warned against the concept of a Jewish nation state in the Middle East. He saw problem upon problem looming ahead and considered this anything but a contemporary, progressive and forward-looking solution. Europe, Steiner said, needed the Jews and the Jews needed Europe. Steiner, unlike many other critics of Zionism, did not mean the assimilation of the Jews as an adaptation to Christian 'Germanness', by adopting German culture and German Christianity. This is precisely what Rudolf Steiner did not mean. Rather, he was of the opinion throughout his life that cultural, artistic, religious content should exist and live in a multicultural coexistence of people, in diversity, respect and regard for each other—and for the diversity of spiritual-cultural life. As I said, also in religious matters. He did not advocate assimilation in the sense of a forced assimilation, but rather in a human way.

Of course, one can say that in 1897, it was in no way clear what waves of hatred and extermination due to anti-Semitic persecution would occur in the coming decades in Europe and especially in Germany—and in hindsight, it is now clear. In the 1890s, Steiner completely underestimated the looming threat for Jewish people. In 1897, he wrote that anti-Semitism was absurd, but would be settled in a short time. Things turned out very differently.

At the end of the nineteenth century in Berlin, he was very active on behalf of Jewish authors in a 'Magazine for Literature' that he edited, so much so that he got into trouble. 'You don't have to be an anti-Semite to dislike publications like this week's issue. This issue gives the impression of being a magazine for Judaism. Jewish questions by Jewish authors, only. For judging by your reputation, you will in most likelihood also be counted among the chosen people,' his publisher wrote to him.[19] However, Steiner did not care; he published the articles by Jewish authors because he thought they were good and important. Around 1900, he began to take a journalistic stand against anti-Semitism. He wrote about the contemporary 'machinations of Jew-hatred', about the 'outrages' and 'degradations' being experienced by Jews. It makes 'no difference' whether someone is Jewish or Germanic, Steiner postulated—anti-Semitism 'poisons' culture. The entire culture was being endangered by anti-Semitism, which corrupted thinking with 'muffled feelings'.[20] In an obituary for one of his Jewish friends, the lyricist Ludwig Jacobowski, Rudolf Steiner formulated in December 1900:

> Old, reactionary powers believed their time had newly arrived. Slogans and dull instincts began to have an effect on the broader masses, which one would have thought not possible for a long time now.[21]

Rudolf Steiner revised this last sentence himself.

In 1897, in his criticism of the Zionism of Theodor Herzl and Max Nordau, he did not believe that anti-Semitism would gain strength again, but three years later, in 1900, he recognized that this was indeed the case; he did not just recognize it by 1919 or 1923 or even 1933, but much earlier, and wrote that 'decisive' and 'unambiguous' statements on the matter were essential.[22] In any case, he did not fail to do so—as we can read in his essays, some of which were written for the 'Verein zur Abwehr des Antisemitismus' [Association for the Defence against Anti-Semitism] and published in its 'Mitteilungen' [Communications], an organizational group in which his friend Jacobowski had collaborated. There isn't one word of all this in the critical publications accusing Rudolf Steiner of anti-Semitism and even 'radical anti-Semitism'. I consider this a deliberate falsification of history, a case of demagogic defamation of a person and his life's work.

Stéphane Hessel, the 93-year-old survivor of Buchenwald, published the book, *Indignez-vous!* in 2010 (translated into German by Michael Kogon, Eugen Kogon's son). In *Time for Outrage,* I find such a treatment of Rudolf Steiner and regarding the factual history, outrageous—I would like to say this in no uncertain terms, at this point and on this day. You don't need to be a 'Steiner fan' to agree on that—in the least that. Someone cannot publicly speak out against discrimination and racism, and then attack, distort and—if you will—'discriminate' against a person in this way, a person who wrote essays against anti-Semitism, spoke out and stood up so clearly against all nationalism and racism and even risked his neck for it, which I would like to discuss briefly later.

Whether Rudolf Steiner was still so critical of Zionism in the 1920s at the end of his life, we do not know. There are indications that he saw what had already started in Israel much more positively and more justifiable by then. He once said about Hugo Bergman, the first rector of the University of Jerusalem—who was very engaged with Rudolf Steiner's anthroposophy and who visited Steiner in Dornach before his emigration—that Bergman had important tasks there. Other pupils of Rudolf Steiner and students of anthroposophy had also done important work in Israel, including the work of Kabbalah researcher Ernst Müller. Rudolf Steiner highly valued the spirituality of Judaism, its entire contribution to the development of human consciousness, culture and science. But he was against the founding of a nation state and against the isolation of the Jewish community, which he saw as a disastrous step backwards, even although it had largely arisen as a gesture of protection resulting from the persecutions.

In this context, he was able to express himself extremely bluntly, especially in his younger years, as well as in the context of his criticism at the time of all known religions and religious claims, moral precepts and norms of behaviour that are not found, understood and internalized by the individual themselves. The practice of isolating individual sentences of his and placing them in an anti-Semitic context can be extremely effective in the media, but is a manipulation technique that distorts the meaning and is extraordinarily dangerous. I would like to add the following clearly, however: I do understand, on the other hand, every person who feels disconcerted and disturbed by Rudolf Steiner's decontextualized statements in these writings, after all that happened after 1933. And I am by no means of the opinion that his formulations in the 1880s were consistently positive, nor of the opinion that he was completely free of anti-Jewish clichés. I have developed this in more detail elsewhere following Ralf Sonnenberg and do not need to repeat it here.[23] But on this 9th of November, I would like to emphasize once again in all clarity: Rudolf Steiner was the opposite of a 'radical' anti-Semite, and was an extremely strong critic of this whole movement.

> I have never been able to view anti-Semitism as anything other than a view indicating that its proponents are inferior in spirit, have poor ethical judgement and are distasteful.[24]
>
> Only the reciprocal effects of individuals should be valued. It makes no difference whether someone is Jewish or German.[25]

Clear Powers of Judgement versus Pseudoscience

However, Rudolf Steiner was not only an opponent and critic of anti-Semitism in the first quarter of the twentieth century, but of racial hygiene, social Darwinism and eugenics as well, which I would like to take a step further. Yes, he was one of the early and far-sighted critics of these threats to humankind and humanity through pseudoscientific developments—these and nothing else constituted these disciplines. At the beginning of the 1920s, they were regarded as widely accepted contemporary science, as 'state of the art' science. Steiner, however, did not tire of pointing out the enormous danger and the absurdity of hereditary genetic degenerative thinking, the 'dogma of hereditary burden' (Steiner) and the catastrophic potential of deriving legislative measures from it (such as sterilization and birth control etc.), indeed of beginning an era of *de facto* biopolitics. As is well known, the first 'eugenics' congress took place in London in 1912, with 700 participants, including renowned scientists, in an enthusiastic, even euphoric atmosphere. Rudolf Steiner saw nothing positive there, other than an impending catastrophe. In 1917, he said of the London Congress:

> Various steps have already been taken in this direction, but as long as they are confined to the realm of theory they can have no deep or lasting influence or significance. It is only when they are translated into practice in the social

order that they exercise a deep and lasting influence. The first half of the present century will scarcely be over before humanity will be faced with an alarming situation—for those who are far-sighted ...[26]

Indeed, the first half of the twentieth century was not over before crimes of sterilization and 'euthanasia', the Nazi murders of the sick and murders of the disabled, including some 10,000 children, had taken place, we know retrospectively. These were not isolated crimes, but they had a paradigmatic background, had been thought through for a long time, had their preceding 'theoretical basis'. In 1919 Steiner said:

> Let the kind of teaching that prevails in our universities continue for another three decades, let social questions be treated as they are now for thirty years more, and you will have a devastated Europe. [...] If in this regard there is not a change in learning, a change in thinking, then a moral deluge will overwhelm Europe![27]

That was indeed the case, as we all know. Steiner's predictions, however, have nothing to do with 'clairvoyance' or 'being clairvoyant', but with his very clear power of judgement for socially relevant tendencies of anthropological science, or rather pseudoscience, which together with the power of the state and support developed into a paradigm that in the end had a destructive effect on humanity. Anthroposophy is not a 'misanthropic ideology' (Sebastiani), rather, it opposed 'misanthropic ideologies' already in the first quarter of the twentieth century, as can be shown in documented detail, and continues to do so to this day.

Now, I would like to add the following: the ideas of racial hygiene and anti-Semitism, social Darwinism and eugenics penetrated from the outer circles of society into its centre soon after the end of the First World War in the context of a completely impoverished central Europe. And I would like to add, that at lightning speed, Rudolf Steiner developed practical counter-models against them with his co-workers. Counter-models: by this I mean the Waldorf schools, the institutions of Anthroposophic Medicine, the first homes of Anthroposophic curative education, the initiatives for social threefolding and economic restructuring, as well as the first farms working with the biodynamic agricultural approach. We can really describe them all as counter-models to the impending 'downfall of the West' described in Oswald Spengler's vision,[28] as well as counter-models to the aforementioned objectives of a biopolitical dictatorship of order that was already then on the horizon. Even the great Vienna 'West-East Congress' of the

General Anthroposophical Society in June 1922 was a kind of reaction to the contemporary, political and socio-political sentiments. A woman who was murdered in Auschwitz in 1944, the Jewish violinist Alice Wengraf, wrote retrospectively, that the Vienna Congress, which she never forgot, had been an atmospheric counterbalance at the time of 'the worst incitement of nations and hostility in the world'.[29]

Here, I don't even want to discuss the fact that many branch leaders, that is, leading personalities in the Anthroposophical Society were Jews: in Stuttgart, but also in the Czech Republic and Italy; I don't want to use that as an argument here at all. There is no need to talk about it at all, because for Rudolf Steiner it was not an argument, but something completely self-evident.

He needed talented, dedicated, intelligent and educated people for the work, and they often came from the Jewish academic intelligentsia and he held them in the highest esteem.

The presence and representation of these people, however, made anthroposophy even more suspect in the eyes of right-wing nationalist and racist anti-Semitic circles than it had been to them before. The propaganda that Steiner was himself a Jew had demonstrably begun as early as 1908, long before the NSDAP, through a Jesuit priest, but it became increasingly massive, especially after the end of the First World War, when Steiner and the anthroposophical movement appeared with the ideas of social threefolding, the overcoming of the nation state and the free Waldorf Schools. The initiative for social threefolding constituted a danger to the existence of the German state and the German 'folk soul', and fighting it was 'a duty', wrote the *Süddeutsche Zeitung*. The *Frankfurter Zeitung* spoke of 'traitors to Germanism'. In a right-wing nationalist magazine it could be read that people like Unger and Ahrenson, members of the Anthroposophical Society leadership in Stuttgart and both Jews, had indicated that a 'spiderweb was surreptitiously being woven over the German people'.[30] The increasing radicalization of this campaign against anthroposophy, in which representatives of the clergy and German university professors had also taken part, has already been thoroughly worked through and published many years ago in a meticulous historical book by Lorenzo Ravagli[31], but as far as I can see, it has not found its way into mass media or the literature of the critics of anthroposophy, is not to be found anywhere. In Switzerland at that time, it was, in particular, the Catholic priest of Arlesheim, Father Max Kully, an avowed nationalist, racist and anti-Semite, who led the propaganda against Steiner and the Goetheanum; he collaborated with the right-wing nationalist and extreme right-wing 'völkisch' circles in Germany. Adolf Hitler's mentor, Dietrich Eckart, was also very concerned about the Waldorf Schools and spoke out about it in his magazine *Auf gut Deutsch* [On Good German] shortly after the school's opening, and especially about Steiner's opening speech. In 1921, in his *Völkischer Beobachter,* Hitler himself spoke out about Steiner as a 'national criminal', a 'gnostic' and a 'supporter of the threefolding of the social organism and all these Jewish methods that destroy the normal mental constitution of human beings'.[32] From the end of 1921, Rudolf Steiner was on an assassination list of right-wing extremist circles. A first attempt failed in May 1922 in Munich, but resulted in the ending of Steiner's public lecturing activities in Germany. The *New York Times* wrote about this assassination attempt: 'Organized reactionaries, nationalists and anti-Semites attended the lecture in force'[33]—while right-wing radical circles in Munich were outraged that Steiner was even allowed to speak in Germany at all without the government intervening. There was talk of a 'Jewish parasite'—'[the father] is said not to have been a Jew, but anyone who has studied the stunted forms of Steiner's thinking will easily recognize that the Talmud and Kabbalah were the sperm from which this strange growth thrived'.[34] I'll spare you the rest; we'll leave that for tonight, but you can experience the atmosphere of hatred and demonization. The wooden Goetheanum, the enormous double-domed building on which people from 17 nations had worked for so many years, even during the World War, was destroyed by arson, in precisely this atmosphere. 'If these gentlemen come to power, my feet can no longer tread on German soil,' said Rudolf Steiner in November 1923 after the 'March on the Feldherrnhalle'[35]; they will bring 'great devastation' upon Europe.[36] Where, I would like to ask, is all this in the writings of the critics of anthroposophy who would have Rudolf Steiner lining up with Adolf Hitler from Austria? Steiner spoke of 'destruction' and of looming evil:

> There are forces at work that give us an inkling of the abysses into which western civilization is still to plunge.[37]

Should one not, a hundred years later, at least be familiar with the words taken from a co-stenographic recording of a lecture given on 1 January 1924? In the autumn of 1924, Hitler was released from prison early, where he had written his basic work, *Mein Kampf.* I will stop at this point with my report of the 1920s.

Anthroposophy and National Socialism

I'm now going to make a leap to 1933. Eight years after Steiner's death, an essay concluding his written life's work appeared posthumously. Called *From Nature to Sub-Nature*, it is an impressive text on how to deal with the forces of evil at work in our time, to be worked through meditatively.[38] I will now begin in 1933 with one of Rudolf Steiner's co-workers and begin for good reason with Ita Wegman, for she was at his side until 30 March 1925 in building up the new medicine, curative education and social therapy. No one witnessed, as she did, how intensively Steiner grappled with the dangers of looming biopolitics and eugenics, elimination based on race and the 'euthanasia' tendencies of her time. Wegman, in addition to her extremely committed work in the building up of Anthroposophic Medicine, curative education, the development of anthroposophical pharmacy and specialist training, continued caring for the socio-political problematic issues.

On 30 January 1931, she had already asked a Stuttgart Waldorf teacher of Jewish origin, Ernst Lehrs, to discuss political questions and 'Social Threefolding' with his upper school pupils. In Berlin, she had noticed the enormous interest of young people in political and social issues, but also that they were being ideologized by both right-wing and left-wing political parties, which were making use of young people, and therefore urgently needed to be opposed.[39] However, doom was drawing ever closer. Two years later, on 17 March 1933, shortly after the NSDAP came to power, Ita Wegman wrote to a colleague in Munich, an anthroposophical doctor, deeply concerned about the 'hateful persecution of dissidents, such as Jews, ...'—'So I would almost urge the doctors to help send the children out of Germany. Remember that we would like to take children in with love here.'[40] Wegman wanted to take children at risk into Switzerland, Jewish children, perhaps even more so handicapped children. I don't know exactly for we found only the letter to her Munich colleague Emmy Giesler in her estate, but not her reply nor any follow-up correspondence on this matter. As far as I know there are not that many letters from doctors of that time (March 1933!) urgently calling for the evacuation of endangered children from Germany—even in the non-anthroposophical sector. Such letters are rare, as far as I know. 'Remember that we would like to take the children in with love here ...' In the same March of 1933, Steiner's co-worker wrote to curative educators in England:

> The conditions in Germany are quite bizarre and extraordinarily difficult to judge, because, in a clever way, the malicious are well disguised

and even the illusion is being created that they are right; an unparalleled seduction is arising there. What has already happened is of tremendous significance.[41]

She saw a 'wave of blind nationalism' gripping so many people who were being dragged along with it. A few weeks later, in one of her letters to a Weleda employee in London, she said:

> It will now probably happen in Germany that freedom will no longer prevail and perhaps even commissions will be set up everywhere to decide on things, both in political life and intellectual life, such as the administration of schools and other things, and such that all Jews will be expelled. That is of course our first concern, our various friends who can no longer stay in Germany, whether they are of Jewish origin or whether they are not completely safe in Germany because of certain work that has taken place, more in the social domain. And for me the anxious question is: how do we organize ourselves as true anthroposophists in order to truly serve humanity, so that we can continue to spread spiritual science in the right way beyond nationalism and also live according to it, because I see it coming that the wave, which now pervades Germany will not only remain in Germany, but will also spread to various other countries and each country will isolate itself until it finally—because this is of course against all true evolution—degenerates into a general great war again. How do we behave—and this belongs to our tasks, otherwise anthroposophy has no meaning at all if we only acquire it for ourselves in hidden closets—towards these major things, in order to work in such a way that we can perhaps prevent some things through our right attitude and the right deeds? [7.4.1933 [42]]

There is still be a lot to be said about this. I would like to see something like this in the *Zeit Online* on 1 September, the anniversary of the beginning of the Second World War, in memory of the doctor, Ita Wegman (1876-1943) and her anthroposophical-humanistic principles. She deserves a tribute. I would also wish for such questions as these in general, on such a commemorative day, independent of Wegman: How can we work in the face of the afflictions of our time? What might we be able to prevent through 'right attitude' and 'right deed'—in Germany and all over the world? These are questions for 1st September, anthroposophical questions, human questions. Anthroposophy has to do with love for the human being. It is the opposite of a 'misanthropic ideology', the exact opposite. It is also the opposite of ideology in general, I also mean.

Anthroposophy as Reflected by National Socialism

I want to stay with the topic of ideology but change the angle of vision. Coming from Hitler himself and in support of Hitler, the national socialists saw anthroposophy as something absurd and dangerous from the very beginning, including of course the state-free Waldorf schools—in a state that aspired to omnipotence, i.e. a totalitarian state. There were early expert opinion statements, as early as the autumn of 1933, against Waldorf schools. According to Professor Hauer of Tübingen, the Waldorf schools were completely incompatible with the national socialist state. He wrote of their 'tremendous danger' to the building of a 'genuine German education' and of many other things.[43] Nevertheless, it took until November 1935 for the Anthroposophical Society to be forbidden, although Himmler and Heydrich had been preparing the ban since the spring of 1934. I have always asked myself, why so late? On the other hand, I have wondered why the leading national socialists took the anthroposophists so seriously at all?

In Germany at that time, there were about 7,000 members of the Anthroposophical Society and perhaps 4,000 anthroposophists in independent working groups, together about 11,000 of mainly non-political people. They did not exactly form a major force in a state that comprised more than 65.3 million people in 1933. Nevertheless, Himmler and Heydrich had comprehensively been preparing the ban. We know that the Nazi regime had initially feigned the legitimacy of the ban as the official regulation still required time. Reinhard Heydrich, later head of the 'Reichssicherheits-hauptamt' (RSHA) and commissioner for the 'Final Solution of the Jewish Question', finally signed the ban on the Anthroposophical Society. In it, the Waldorf schools were explicitly named as a danger because they pursued an 'education oriented towards the individual' which had nothing in common with the 'educational principles of national socialism'. The wording of the ban also spoke of the great contrast between the Anthroposophical Society and national socialism by virtue of anthroposophy's 'internationalism', 'pacifism' and its close factual relationship with Jews.

> As a result of this opposition to the National Socialistic idea of Volk [Voelkische Gedanke], the continued activity of the Anthroposophical Society imposes a danger of harming the National Socialistic State. The organization is therefore to be dissolved on account of its subversive character and the danger it poses to the public.[44]

House searches and confiscations followed. A 'community of people based on individuality' (Steiner [45]) was not tolerated in Hitler's state, and one of the many SD reports from the main office in Berlin in May 1936—that is from the SS Security Service under Heydrich—states:

> [...] Anthroposophy detaches the spirit from its connection with race and the 'Volk' and condemns the racial and the nationalistic to a lower sphere of primitiveness, of instinct and drive from prehistoric times to be overcome by the spirit. It thus proves that it is intertwined with the main currents of European intellectual history to date, above all the Enlightenment, German idealism and liberalism of past centuries. [46]

This 'intertwining' did indeed exist and other reports and accusations were also mostly true. It is true—as the reports contend—that there was no convergence whatsoever in the content of anthroposophy with the national socialistic world view; that the Goetheanum in Dornach was built by people from 17 nations being a 'precursor to the building of the League of Nations'; that the Waldorf School with its 'education of the human being' was without any 'völkische Bindung' [bonding to the 'Volk'] and orientation to authority; and that it only adapted 'formally' to the new regulations and the new spirit etc. [47]

In his expert opinion on the Waldorf schools, the regime's top political educator and only senior with any knowledge of anthroposophy, Professor Alfred Baeumler, wrote that in Steiner's anthroposophy there is no 'biological-racial' thinking whatsoever, no concept of 'völkisch' community; Steiner substitutes 'humanity' for the 'Volk' and had virtually developed the opposite of a biological doctrine of the human being:

> 'The fateful distinction', he wrote, 'occurs through the fact that Steiner replaces the theory of heredity with a *different, positive theory*. Steiner does not simply overlook the biological reality, but rather consciously transforms it into its opposite. Anthroposophy is one of the most consequent antibiological systems.' [Professor Alfred Baeumler] [48]

Baeumler indeed saw this correctly. 'The anthroposophists are the worst enemies of our Reich,' wrote Professor Hauer to Professor Karl Astel of the 'Institute for Human Hereditary Research and Racial Policy' at the University of Jena and head of the 'Thuringian State Office for Racial Affairs' and judge at the Hereditary Health Court in Jena. [49] Anthroposophy 'will ultimately lead to the disintegration of the national socialistic worldview' and pose a considerable danger to the 'unified orientation of the German people', although quite consciously, the anthroposophists themselves display an 'appearance of being harmless', written in a detailed report of the Reich Security Main Office as late as 1941.[50]

Even Reinhard Heydrich, by now head of the powerful RSHA in Berlin's Prinz-Albrecht-Strasse with a total of about 3,000 employees, still spoke of anthroposophy as a 'special doctrine endangering national socialism', a 'dangerous factor' and of the 'anthroposophical contamination of the Volk' only six months before his death.[51] He obviously took anthroposophy as seriously as Freisler later did with regard to the numerically very small anthroposophical group, the Kreisau Circle under Helmuth James Graf von Moltke. And again I would like to ask: Where does this information actually appear in today's reports, why is it never mentioned when anthroposophy is reported in relation to national socialism? Surely such testimonies are not arbitrary or irrelevant; for, if you like, anthroposophy was described by the highest authorities for what it was and is—a radical opposition to Nazi ideology.

Why were some anthroposophical institutions able to continue working after 1935?

Having arrived here, however, I would also like to raise the question once again as to why anthroposophy had not been completely banned as early as 1933 and why all Waldorf schools had not immediately been shut down, since, from the outset, the regime was not exactly circumspect in 'dealing with' groups whose 'character was beyond doubt deemed a threat to the state'? However, I think it has to be taken into account that the anthroposophists did not form a political party, did not express themselves politically as a rule, and were one of the countless 'world view' communities of people who met, studied texts and were harmless. Without public, state-recognized private schools—of which there were an enormous number in the German Reich, including Waldorf Schools—and without an imposing and 'questionable' central building in Switzerland, whose significance as the 'centre of the anthroposophical world movement' was difficult to assess, the national socialists would probably have paid as little attention to the anthroposophists for years as they did to other small groupings with a few thousand people who were, for instance, buddhistic or nutritional groups, who cared about vegetarian food, or the deities of the Indian people. However, the schools of the anthroposophists, whose principles had nothing to do with 'national socialist educational principles', were a bone of contention, even though they had comparatively few pupils in relation to the whole of Germany. And the closure of private schools generally caused certain problems for them.

In official statements, the representatives of the anthroposophical schools defended themselves to the best of their abilities. Under pressure, they joined, at least in part, the National Socialist Teachers' Association; they claimed to be 'for Germany' as well, for the land of Goethe and 'Goetheanism', for German culture and the 'folk soul', etc. Statements were sent by the Waldorf schools to Berlin and to the ministries of culture, full of rhetoric about their adjustment, intended to signal a willingness to compromise, be quickly transparent and to formally express an attitude not of resistance. The Anthroposophical Society's leadership in Dornach also submitted corresponding papers: that Steiner himself had been a good German, and he too had voted against the Treaty of Versailles, against Western 'intellectualism', etc. These submissions were to pay lip service to the regime, but as we now know and can document, were highly controversial internally, among the teachers of the Waldorf schools and among the Society's Executive Council members, with people like Ita Wegman and Elisabeth Vreede, who were however unable to prevail. Those who 'adapted' and took an 'opportunistic approach' temporarily determined the playing field and saw their approach as strategic, and as one would say today, as having 'no other alternatives'. The German National Executive of the Society wanted to close the Anthroposophical Society in Germany of their own accord in 1933, rather than accept any form of 'cooperation' or 'adaptation'. The treasurer in Dornach, however, would not allow this as he was dependent on the membership income from Germany for the Goetheanum. People were struggling for their own survival and that they fought against the threatened ban and closure of the Waldorf schools, showed a great—in the least, rhetorical—willingness to compromise, hoping for some kind of 'coexistence' while living under pressure and in a great deal of fear.

Most members of the General Anthroposophical Society went into an 'inner hibernation' as early as 1933, when the hate attacks against Steiner were widely reported in the daily papers. It was better, they thought, not to be publicly known as anthroposophists in a state that increasingly, year by year, spied on people, using an array of surveillance and denunciation strategies. How many anthroposophists nevertheless sympathized with the Nazi regime, at least in its early 'idealistic' phase, and despite the hateful attacks against Rudolf Steiner, we do not know—being of a Society that united people of all political persuasions. Someone certainly did not come to national socialism out of anthroposophical thinking, but possibly as a private person, in spite of anthroposophy—I will come back to this point later. Finally, there was a small but energetic group of anthroposophical activists who had been in talks for years with senior Nazi officials, from whom they learnt that Nazis valued some practical applications of anthroposophy: from Weleda remedies (in the medical and cosmetic fields) to the Demeter products and some aspects of Waldorf education.

Working Together with National Socialists

Uwe Werner worked in detail through the activities of these negotiators and their contacts in the oligarchic power apparatus of the Nazi regime more than two decades ago and published them in a large 488 page book.[52] The negotiating skills of a personality such as the Dresden Waldorf school teacher and de facto headmistress, Elisabeth Klein, and her connections to people on Rudolf Hess' staff, including Alfred Baeumler and a high official in the Ministry of the Interior, Lotar Eickhoff, are to thank for the continued existence of the 'free' Waldorf schools in the Nazi state for many years, far more than the rhetorical appeals of numerous school representatives achieved. Yes, Baeumler—who himself radically rejected anthroposophy as such, but found some aspects of Steiner's philosophical and Goethean writings, and elements of Waldorf education to be good and useful, after he thoroughly examined them (and possibly having a 'handicapped' child in need of special care himself[53]), and even listened to and showed an interest in Elisabeth Klein. Baeumler visited Klein's school and was impressed by its educational results and by the appearance of the pupils as well. He came to the conclusion that the epochal teaching method of the Waldorf School was good, as were many other methodological and didactic aspects, including the relationship of the teachers to the pupils; however, the schools would only have to introduce the racial concept of Nazi biology, teach history differently, consistently leave out anthroposophy and employ politically 'reliable' teachers—only then could they not only continue to exist in the Nazi state, but be a significant model. Baeumler therefore became a protector of

Waldorf schools without anthroposophy and Klein ostensibly went along with it. In Dresden, however, they did not follow what Baeumler wanted and demanded. I will however come back to this later, but you may notice that we are in a 'grey zone' here.

The case of Otto Ohlendorf, the high SS man from the RSHA, is also very well known. Ohlendorf stated after 1945, very probably correctly, that he himself had never had anything to do with anthroposophy, but that he had refused to close down and destroy 'living, constructive institutions and research' in Germany, especially as national socialism had in the 1930s not yet succeeded in achieving its own independent 'shaping' in areas of life such as organic agriculture or in the production of remedies. He had been interested in 'utilizing' existing work 'in the interests of Germany and national socialism'.[54] As with Baeumler, Ohlendorf was interested in its practical application, despite the anthroposophical basis of ideas, 'in the interests of Germany', which for him were synonymous with the interests of national socialism. We know today that the SS succeeded in employing biodynamic experts from the Demeter association on their own medicinal plant farms as staff, even at the Dachau concentration camp. People from anthroposophy who contributed their professional expertise, who apparently had no scruples, were hoping for personal survival or professional advancement and very likely for the advancement of biodynamic agriculture in general as well. Thus began a highly problematic collaboration, albeit with only a handful of people, who made themselves available to national socialism and the institutions of the SS.

However, this can only be said of the employees of the SS medicinal plant facilities in the field of biodynamic agriculture, but not of Elisabeth Klein and her Dresden Waldorf School. Although the school was able to continue to exist as the only Waldorf school until 1941, due to its protectionist backing and its status as a state 'model school', before it was closed and Klein arrested. Klein consistently refused to change the teaching staff (with the recruitment of 'politically reliable' teachers) or the teaching content, even for the teaching of history, throughout these years. Despite Baeumler, there were no 'racial studies' at the Dresden Waldorf School up to its closure in 1941. Klein also never joined the NSDAP, nor the National Socialist Teachers' Association. Her dangerous and in various respects problematic path—which I have examined and discussed in somewhat greater detail within a larger book in 2019,[55]—was of a composite mixture of anthroposophy and national socialism, despite its shadowy sides; the idea of coexistence of both directions or of 'free' schools existing in a totalitarian state, was a profound illusion, as Karen Priestman, among others, has pointed out.[56] All in all—and there are historically detailed and in-depth studies on this by Norbert Deuchert, Wenzel Götte, Uwe Werner, Detlef Hardorp and Volker Frielingsdorf, among others—it is true that the Waldorf school outwardly went the route of a 'rhetorical adaptation' but inwardly, continued with its own high ideals and educational continuity. This is not an assertion, but is extensively well documented, among other things by the many critical reports of the Nazi school inspectors, which have subsequently been evaluated by Deuchert and especially Götte in his dissertation. The inspectors consistently criticized, even expressed horror at the fact that the content and form of national socialistic reality had been bypassed without even a trace in the Waldorf school.[57]

What was true for the Waldorf schools was also true for other anthroposophical institutions, as long as they were able to continue working through protection or favourable circumstances, some of them until the end of the war. Within the framework of our scientific study of Anthroposophic Medicine in the Nazi era, we were able to evaluate not only 125 medical reports with around 800 individual case histories on the efficacy of anthroposophic remedies in 1941/42, but also over 1,400 case histories from the Wiesneck Sanatorium, a psychiatric and general anthroposophical clinic, during the Nazi period. I would like to say that one can only take one's hat off to the way the anthroposophical clinic worked during the persecution of mentally ill people, held its own in the midst of adverse conditions and devotedly continued its humanistic psychiatry and psychotherapy, its individual patient care, as a real counter-model to the 'inhuman ideology' of the Nazi era and often enough, of its nearby Freiburg surroundings. At this point, I would also very much like to say something about the impressive continued work of the curative homes with an anthroposophical orientation, which diametrically opposed the Nazi aims of registration, segregation, sterilization and 'euthanasia', and which continued to support and protect the children and young people, but I cannot do full justice to this here due to time constraints. Perhaps I may nevertheless, *pars pro toto*, briefly mention a fairly unknown person, the anthroposophical curative educator, Hubert Bollig, with his small home in Malsch near Karlsruhe, which was opened in 1931. After considerable difficulties with the local Nazi authorities, Bollig was ultimately forced to close in 1939. He managed to safely house and care for 33 of the 40 children and adolescents until the year of the start of the euthanizing of children known as the T4 action. He set off on a journey across the Black Forest to Lake Constance with the seven remaining children in search of places of protection for them, which could not be recognized as institutions. Later in 1940, surprisingly, he was able to return and began curative education work again in 'Waldhaus' in Malsch before he was taken into Gestapo custody. Before that, however, he again managed to put children into safe custody, including abroad. This has been historically reappraised by non-anthroposophists. With the exception of one boy, Otto Nikolai, Bollig succeeded in placing all of them safely, with the exception of one Jewish child with Down syndrome whose father had fled to Chile and older sister committed suicide. Eventually, however, Bollig managed to take Otto into his own family by declaring him to be indispensable to his wife, who had a walking impairment. Otto Nikolai survived the Nazi regime and stayed with the Bolligs, who opened their home in Malsch for a third time in 1948.

Of course, it can be said that these are individual cases and that these existed everywhere. Regardless of whether there are other 'cases', or people with an individual, unique life and destiny, I think that Hubert Bollig's story is remarkable. A curative educational commitment like his in those difficult and dangerous times should be recognized. Aleida Assmann wrote, not with regard to anthroposophy and anthroposophists:

> The people who helped those persecuted by the Nazi regime were a small minority, but it is all the more important that they are not completely forgotten. They were not themselves concerned about being remembered. [...] These helpers serve as role models and are an indication that even in a 'genocidal society' there were still opportunities for making personal decisions and finding scope for action.[58]

Regarding Hubert Bollig, I would like to add that he did not only act on the basis of his personal morality and decency in the midst of a 'genocidal society' (Assmann), but on the basis of an ethics that had arisen in him as a consequence of anthroposophy's humanistic image of humanity, such as its anti-racism and anti-eugenics positions. So, I would ask you to consider this. For him as an anthroposophist, *every* human being is an I-bearer and an I-being, even when the forms and difficulties of expression of their individuality are different. Because of his anthroposophical image of humanity, Bollig was an anti-racist and anti-eugenicist and a radical opponent of so-called 'euthanasia'.

Perhaps I may also briefly quote another anthroposophical curative educator in this context, the 29-year-old Werner Pache, who wrote to Ita Wegman, from Berlin to Arlesheim, on 12 February 1933, two days after Hitler's first 'Sportpalast' speech as Reich Chancellor, which Pache had heard. He asked Ita Wegman to visit the curative education home Schloss Hamborn near Hamborn as soon as possible and wrote this to her:

> Two evenings ago, when I stood on Wittenberg Platz in Berlin, where, as in many other public places, Hitler's speech resounded through loudspeakers and the power of Ahriman was almost overwhelming, I was suddenly imbued with a very confident, calm feeling, which was so strong that I felt suddenly released from the spell of this evil power. A feeling that our cause will not perish.[59]

By the 'power of Ahriman', Pache meant the power of evil.

Self-critical Questions for the Anthroposophical Society

I now come to the last part of my talk on this 9th of November. And I would like to emphasize: that despite all of this, despite Wegman, Bollig, Pache and many others, we have to ask critical questions, of history and of ourselves. It is anything but sufficient to reject outrageous allegations and suggestive, manipulative media reports—and then just sit back. On 9 November, it is much more appropriate to critically question the anthroposophists of that time, but especially ourselves as well. Yes, and you? Have you done enough, have you really done enough, are you doing enough? Did the General Anthroposophical Society react vigilantly enough in 1933–1945; did it do everything to protect people in danger? Did it commit itself enough, against national socialism and for a different, humane society? Did it really follow the path of Ita Wegman, who demonstrably helped a great many refugees from Germany?

Ita Wegman, this wise woman, once wrote (in one of her many letters): 'As long as a teaching, remains only a teaching, no difficulties arise.'[60] This can mean in the context we are considering here: anthroposophy and the anthroposophical image of man is indeed impressive, great, humanistic, anti-racist and so on, included within it, the teaching content of anthroposophic medicine, education and curative education. But to live something, to put it into practice—especially under the conditions of an authoritarian or totalitarian system, at any rate under the conditions of an 'existing' social system—it is not very easy, as history past and present shows. The other thing I derive from Wegman's sentence and would like to emphasize is this: the teaching is a teaching; it is a body of ideas that we are able to internalize and which can change us.

The Christian can internalize the Sermon on the Mount as a body of teachings and become a different person, pursue different goals, behave differently through it. The Jew or Muslim can also practice this in existential experience, through reading, praying and meditating on their sacred texts, which often carry a high morality within them, high moral aspirations. However, this internalization of the ideal and the path to an ethical morality can, as we know, also be omitted, the ideas can remain in the head and the rest goes unchanged. Leading national socialists loved culture, but acted barbarically, violently, cruelly and brutally to the highest degree. Therefore, without wanting to compare this in any way, but following Ita Wegman's statement, we need to—I mean in critical contemplation—consider that for many anthroposophists the 'teachings' of anthroposophy remain mere 'teachings', although they bear within them an invitation to an inner training and self-transformation. And because this was very possibly the case, quite a few anthroposophists probably remained elitist and arrogant, nationally conservative, patriotic or anti-Semitic in their 'other' lives and other personalities, despite Steiner and anthroposophy, like so many other people, including the many Christians, despite the Sermon on the Mount.

How far anthroposophy transforms people, or remains 'content of the head', is left to the freedom of the individual. Anthroposophy can remain head content and the human being continues living in their own usual way. In my opinion, it would be completely wrong to claim that the approximately 7,000 members of the Anthroposophical Society in Germany, for example, were all imbued with the ideas of social threefolding, were extremely critical of the centralized nation-state and observed the initiatives of right-wing radical forces with the same suspicious eye as Rudolf Steiner had done, at least after 1923. This was certainly not the case and it is proven that some anthroposophists were of the opinion, at least in the first years after 1933, that something very great had come in Hitler, wrote adoring letters to the 'Führer and Reich Chancellor', as did so many 'ordinary Germans'—but they did it as well. In one letter Ita Wegman spoke of a 'tremendous seduction' taking place and was shocked that not all anthroposophists were seeing through the seduction.

And we should not forget that much can be made out of anthroposophy, or rather out of Rudolf Steiner's hundreds of volumes of lectures. 'Anyone can saddle my teachings to their own head', Paracelsus once complained,[61] and we know countless examples of the misuse of Rudolf Steiner's lecture content and passages from his writings by people who considered themselves anthroposophists, right up to Werner Haverbeck and his wife. One can integrate passages of Rudolf Steiner on Germany into one's own German nationalism or German imperialism, or use Steiner's critical remarks on Zionism as the 'anthroposophical' basis for justifying one's own anti-Semitism, and so on. It is limitless and these problems extend into our present time—even though Rudolf Steiner himself had a very clear and well-defined position, which I tried pointing out at the beginning.

Repressed Questions in Dealing with Nazi Protectors

How the Anthroposophical Society as a whole behaved from 1933 to 1945 and what its members thought is not easy to research; we have been trying to do this for some years now for the more than 600 anthroposophical doctors of that time. I would like to say, the preliminary results are overall very positive with regard to their resistance to fascism and its biopolitics, to its ideological and structural 'conformity' as well—and this is not surprising if one knows Rudolf Steiner's *Course for Doctors*, his medical textbook with Ita Wegman, his *Curative Education Course* and the whole direction that was thereby established and to which these doctors and therapists had committed themselves, with regard to medical ethics and morals, and the morals of curative education, as well. And yet there were some outliers, conformists and collaborators: 'As long as a teaching remains only a teaching, no difficulties arise ...'. Of course, not all them were heroes and martyrs, and obviously not all of them had the stature of Wegman or Bollig, I mean, their inner stature, their uprightness, as we probably don't have either.

Obviously, non-collaboration as such is not in itself a predicate, a special award or a unique attribute. For often an intentional counterpart was also missing, by which I mean, who among the national socialists wanted to 'collaborate' with anthroposophists, unless they were excellent representatives of biodynamic agriculture or otherwise had something very unique to offer that was of special interest to the SS? Even anthroposophists who might have wanted to become NSDAP members did not find it easy to be accepted. Later, some claimed that anthroposophy per se made them 'immune' to national socialism. What arrogance—and how little historical support exists for this claim.

In my opinion, since the 1980s and beyond, this arrogance has generated some backlash, very understandable reactions, in my opinion. And of course, it is also legitimate to ask why there are still no critical biographies on the collaborators from within our own circles? Why has nothing been published from the anthroposophical side about Sigmund Rascher—or about his father and uncle, Hanns and Fritz Rascher, both of whom were anthroposophical doctors and close to the Nazi regime? Hanns Rascher was even an employee of the SS security service or was, at a minimum, an informant. In the obituary of Hanns Rascher by Wilhelm Spieß, in 1952, there is not one word about this, but only the high esteem in which he was held for his medical activities. Of course, such concealment and suppression was not unusual in Germany at the beginning of the 1950s, but the rule seems to apply: out of 'loyalty' and collegial cover, or, a misunderstood commemoration of the dead, in which only positive things were to be spoken of. Anthroposophists who worked for the SS in Dachau, such as the botanist Franz Lippert or the biologist Martha Künzel, were also not spoken of in anthroposophical circles for a long time, apparently because it was too embarrassing.

Furthermore, it must be self-critically questioned how anthroposophists actually dealt with their 'protectors' after 1945, with people like Baeumler and Ohlendorf. The autobiographies of Elisabeth Klein, Wilhelm zur Linden and Rudolf Hauschka, in the 1960s, skip over the crimes of these people, especially Otto Ohlendorf, in an extremely disconcerting, even shocking way, although the cruel deeds of Ohlendorf and the Einsatzgruppe D led by him had already been made public by the Einsatzgruppen trial in the late 1940s. Here too, Klein, zur Linden and Hauschka were no different from many Germans who, despite the trial testimonies and the undeniable facts, maintained the myth of the 'good' Otto Ohlendorf, that he had been forcibly transferred to Russia by Heydrich and Himmler and had protected human lives there, suffered morally terribly and made the situation as 'humane' as was possible within the situation.

Klein, zur Linden and Hauschka unconditionally, uncritically and completely followed Ohlendorf's self-portrayal and stood up for him.[62] Did they ever take note of the other side of his character and work, and were they interested in the inconceivable suffering he had been a part of causing and for which he was held jointly responsible? And even after 1945, they mentioned only the support they had received from him for their own endeavours. Did they feel an obligation of loyalty and feel bound to him? All three of them—Klein, zur Linden and Hauschka—wrote in a very self-confident tone, without any detectable self-doubt or self-reflection with regard to Ohlendorf, and with regard to their 'loyal' relationship to him. Are any of them then surprised if this makes people suspicious that they are also suspect, and with them, the whole of anthroposophy? Was the Dresden Waldorf School the only important issue and did the more than 90,000 victims of Einsatzgruppe D under Ohlendorf in Ukraine and Russia not count, nor the Jewish villages that were wiped out, including, among so many others, the destruction in Czernowitz, the birthplace of Paul Celan? Can I really be surprised if suspicion has arisen in the public, a suspicion that something went wrong with anthroposophy and anthroposophists in relation to national socialism?

I also think we have to ask ourselves, were there ever sufficient thought and conversation in anthroposophical circles and in social contexts after 1945 about the break in civilization caused by Nazism? Is it even possible to continue speaking as before, with the same words and concepts? Just continuing to use them after all that has happened—'Germany', 'the folk soul' and so on? And besides, what about political consciousness? Klein, zur Linden and Hauschka obviously had no political-social consciousness in the 1960s, at least not with regard to Ohlendorf, Baeumler and national socialism, no consciousness of the force, brutality and the system structures in which and for which Baeumler and Ohlendorf had worked. They wrote their autobiographies out of their own human experiences and encounters, out of their own 'I', without any apparent contextual awareness. They relied on their personal impressions and on the experience of 'the good in the other' and did not go beyond that, and at no point relativized themselves or the situation. Is that sufficient—given what the RSHA was, what the 'Einsatzgruppen' were, and what Ohlendorf's role in it all was? Or is this still a manifestation of a certain 'inner emigration'—a drawing on the victim role? The Anthroposophical Society had been forbidden in November 1935, and in June 1941, it was closed down even further. However, is it enough to make global use of a victim status and the victim narrative, a narrative that focuses exclusively on one's own suffering—as seen in the autobiographies of Klein, Hauschka and zur Linden—and which externalizes all guilt without asking whether one acted adequately oneself? Aleida

Assmann has analyzed this recourse to the 'victim role' very well, even if not with regard to the Anthroposophical Society. On 9 November, we can ask ourselves whether there are not tasks for us, tasks of a differentiated appraisal of our own fellow-travellers, of the dangers of 'inner emigration', of the overriding concern for the survival of our own institutions and of the seduction of protection and collaboration. One cannot claim the status of being part of the resistance to a 'system of thought', when one never was. 'Humanity is expected to face the truth' (or, 'The truth is reasonable for humanity'), said Ingeborg Bachmann. Her sentence is also inscribed on the memorial stele for the murdered Alfred Herrhausen.

Where does the hatred for anthroposophy come from?

As I already said in my introduction to this evening, actually, we have to work on these questions, which are always present and future ones as well, not to criticize and pillory Klein, zur Linden and Hauschka, but in order to do things differently and better. But that is precisely what is currently more than difficult to almost impossible in view of the vehemence of the current defamatory attacks. One is forced into the role of apologizing, although something quite different is necessary. And finally, I would like to ask the question once again: Where does this journalistic hatred against anthroposophy and anthroposophists come from? Is this just a backlash against anthroposophical idealism and pseudo-idealism, which claims to have been 'immune' qua anthroposophical spiritual science? And then it is also revealed that the anthroposophists were not all that. But does that explain everything? Is it therefore necessary to stylize Steiner as a radical anti-Semite and racist and inhuman ideologue, which he definitely was not, however unusual his spiritual work may be?

What are the driving forces in this aggressive fundamentalism in the fight against anthroposophy? Is it because since the 1970s—at the latest—modernity became equated with things like secularization, atheism, positivism, technology and urbanization, and any social group that represents other points of view and has a charismatic founding personality with epistemological appeal and is esteemed internally, is suspected of being an arch-reactionary, if not fascist, at a minimum quasi fascist, regardless of its conduct—or partial misconduct from 1933-1945? Is it? Peter Bierl claimed in 1999 that anthroposophists were always in the process of 'preparing authoritarian to fascist forms'[63]. How does someone come to such claims—in view of the concrete living reality of anthroposophical initiatives in medicine, education, curative education, social work, farming, economics, art and religion? Is this malignant criticism rooted in the fact that some of the fundamentalist critics of anthroposophy create their own professional identity by 'exposing cults'? Ravagli once wrote that they would have to give up their own identity if they were to acquire the spiritual perspectives of anthroposophy, even as an experiment. 'They would have to give up their habits of thought, their styles of thinking, a result of their education, and they would have to free themselves of economic dependencies, just like the religious sect commissioners who constantly have to find new religious sects because they would otherwise be without work.'[64]

On the other hand, there is also fundamentalism in anthroposophical circles, including the tendency to present all and any of Rudolf Steiner's statements per se as results of his spiritual research and to treat them as 'truth', and this even in situations where Steiner narrated passages in lectures, told anecdotal stories or made humorous comparisons, which by no means all succeeded. Steiner too had personal opinions and moods, sometimes even prejudices; he was a human being among human beings. 'After all, we all judge from a personally coloured point of view, to which the place and time of our birth has brought us'[65]—he said so himself and never denied it. Nevertheless, he was neither an Austrian German nationalist, nor a racist, nor a 'radical anti-Semite'. Who, if not we, who know his work and his biography more intimately, can perform the task of differentiating more precisely, through contextualization and problematization?

This work, however, needs a certain positive attitude of receptivity within the Anthroposophical Society and in the public—and is not at all feasible in a militant atmosphere of aggression, and if at all, only privately, in which case, it is nonetheless not socially effective. As far as the public is concerned, I consider as a minimal requirement that official journalism acknowledges that Steiner created an impressive, epoch-making work—in socio-political terms, he saw and described, like few other people, the anti-Semitic, social Darwinism, eugenics, right-wing radical and totalitarian dangers, in the first quarter of the twentieth century, as well as the socio-political danger of a paradigmatic body of science and its dogmatic determinations, and tried to launch model initiatives with completely different orientations. Furthermore, that his students and co-workers have done remarkable humanistic work, even during the Nazi era. For me, this would be the least and most basic prerequisite for a differentiated discourse on the weaknesses of the anthroposophical movement and the hermeneutic problems of Steiner's work and its reception. However, that is not what it looks like at the moment. I would say, in fact, it is quite the opposite. And so I would like to conclude by repeating the question: Why is this so?; even independent of the aggressive critics of anthroposophy, who have always existed with varying angles of attack: in the first quarter of the twentieth century from the right-wing nationalist camp and in the last quarter from part of the left-wing green socialist camp, as Lorenzo Ravagli pointed out.[66] And now, why these reports in *Zeit Online* and in so many other regional and national newspapers? Of course there have always been benevolent or balanced accounts and critiques, especially recently regarding *Waldorf 100*. Nonetheless, the extent of the denigration and aggressive mockery is astonishing, and it has increased in recent months, undoubtedly.

I read with interest the book by Professor Rainer Mausfeld, who held the Kiel Chair for Perception and Cognition Research for a long time, *Why do the lambs remain silent? How elite democracy and neoliberalism is destroying our society and the foundations of our existence* (Frankfurt 2019). In it, Mausfeld describes, among other things, that the so-called 'western community of shared values' as an established economic and power order is supported by mass media and works by means of selective consciousness control; according to him, mass media stabilizes to no small extent the social and economic status of those who 'own' them or on whom they are economically dependent. They do this through selection and interpretation of 'facts' and control, according to Mausfeld, actual 'opinion and outrage management' that works with a guided manipulation of attention and fervour, with psychological strategies of manipulation of consciousness and subconscious indoctrination. According to Mausfeld, this is the only way to achieve population-wide majorities for wars of aggression in the guise of 'humanitarian interventions'. Mausfeld speaks about 'invisible ideologies' and their unquestioned framework, about the illusions of freedom and democracy that are at play, and about the authoritarian ingraining that occurs, mostly unnoticed. He describes techniques of 'fragmentation' of originally sensible things or meaningful contexts, and the system of so-called 'experts' being utilized as a real instrument of power. They spread an 'illusion of being knowledgeable' (Paul Lazarsfeld) through decontextualized, isolated 'bites', especially among the bourgeois 'educated' classes—I am reminded of the bestseller biographies of Steiner by Helmut Zander and Miriam Gebhardt—and pursue rapid popularization of their views, which they achieve through a radical reduction in complexity and constant repetition of the same accusations and idiosyncratic interpretations. In the process, fundamentally different views and groupings are excluded and portrayed as being basically incapable of discourse:

> Defining what makes up the acceptable, still-reasonable region on the spectrum of views is therefore an effective way to manipulate public opinion: [...] Having the power to control what counts as the boundary between 'still-reasonable fringe ideas' and 'unacceptable extreme views' in the publicly visible range of opinions thus goes a long way in managing public attitudes. [Mausfeld] [67]

Authors or groupings that evolve important perspectives and question established power or the status of the establishment, that are portrayed as not worthy of discussion or of being taken seriously, and as undesirable 'emancipatory movements' that question the neoliberal system of rule and its intellectual elites, are, according to Mausfeld, marginalized and mercilessly defamed. Everything serves to stabilize the status quo and suppress or minimize the social need for alternatives, in a worrying homogenization and subliminal ideologization that itself works with explicit labels of ideology critique:

> The ideological framework narratives are now so deeply anchored in our culture that we no longer even notice them as ideological elements. [Mausfeld] [68]

When I read Rainer Mausfeld's analyses—originally in a completely different context—I was reminded of the widespread media treatment of Rudolf Steiner and anthroposophy. Justus Wittich wrote 21 years ago, after recent accusations of 'racism' against Rudolf Steiner:

> The deliberate aim is to damage the [...] reputation of Rudolf Steiner and anthroposophy: whether this involves alleging Satanism or 'racism', the preparation of the Holocaust, Semitism or anti-Semitism, or membership of a decadent Masonic order, is quite irrelevant; obviously it is not about seeking the truth, but [about] the attitude: the main purpose of which is that something sticks. [69]

Anthroposophy can become a 'target of outrage management' in Mausfeld's sense—as it became so early on. Steiner stated in a pointed description of the situation as early as 16 February 1923, six weeks after the

Goetheanum was destroyed by fire as a result of massive hate propaganda with a destructive intention:

> Opponents tear all sorts of things out of the writings, interpret them most absurdly, spread them with the intent of causing outrage, so that anthroposophy becomes well known, but caricatured by the opponent.[70]

Steiner's co-worker, Louis Werbeck[71], wrote of a 'planned systematic manipulation of public consciousness', and was the first person to describe the total extent of aggressive publicity against anthroposophy in a two-volume work (1924). 'Much of present day public opinion derives from the fact that people are so terribly superficial and pay no attention to the facts' (R. Steiner[72]) was already interesting then and still is today. Rainer Mausfeld quoted Franco Basalgia and what he described Mausfeld pursues in the realm of mass media journalism, which leads me back to my starting point of the report in *Zeit Online*:

> It is high time to speak not only of the great wars, but of the little wars as well, that ravage in everyday life and which know no truce: in peace times, before war, these are the weapons, the instruments of torture and crime, which slowly lead us to an acceptance of violence and cruelty as a normal state in life ...[73]

'Violence' and 'cruelty', the 'little wars in everyday life', can also be represented in the caricatures of ridicule, humiliation and mockery, discrimination as the modus operandi of condemnation—and its target can be anthroposophy, with its innovative, potentially system-changing potential, with its conception of a humane society that asks uncomfortable questions and has something challenging about it. Anthroposophy, enables an 'emancipatory movement' in the sense of Mausfeld, and its counterforces, and therefore the forces of inertia, are massive, even although countless people today know that we have long been in a global systemic crisis, in medical, social, economic and ecological terms. In my opinion, anthroposophy has a significant contribution to make towards a fundamental change. But when it is publicly associated with anti-Semitism and racism, and these labels have been attached to it, it gets thrust out of public discourse and rendered invalid. This does not happen by chance or as a result of a mere chain of unfavourable circumstances. This view of mine does not make me a 'conspiracy theorist', I would like to pre-emptively underscore outright, when I reject the next wave of discrimination.

In view of this state of affairs, we have to, in my opinion, represent anthroposophy proactively and bring it into the light, not allowing ourselves to be pushed into corners which have nothing to do with us, not exhausting ourselves through endless reactionary and defensive actions. Rather, we have to bring our positive contributions to the public, show the 'essential character of the whole' (Steiner's) and our concrete work in the endangered areas of civilization, while at the same time, remaining self-critical. These are contributions and attempts, nothing more; rather small contributions and attempts, but which are, nevertheless, about creating counter-models helping against the 'downfall of the Occident' and the whole world, which is still ongoing, even one hundred years, after Steiner versus Spengler. It is about civil courage in the face of not inconsiderable totalitarian dangers, which have continued to persist, even after 1945, as Alexander Mitscherlich analyzed, early and correctly[74], which are currently and unquestionably booming.

Anthroposophists in the Concentration Camps

9 November. 'One evening, when the sun had gone down and not only the sun, ...', Paul Celan begins his prose in his *Conversation in the Mountains*.[75] The sun, and not only the sun, began setting on 9 November 1938, or had long since set on the spiritual-cultural, human, scientific and religious sun of Judaism in Germany. Many Jewish anthroposophists were eventually also sent away to camps and perished. *The Drowned and the Saved* is the title of Primo Levi's last, extraordinarily lucid book, his profound reflection on the system of the camps. From the ranks of the Anthroposophical Society in Prague alone, some 70 Jews were deported to Theresienstadt. We know from one survivor that most of them took with them not only the Bible, the Old Testament, but also one of Rudolf Steiner's writings. We know that there was an underground anthroposophical network in Theresienstadt, a working meeting for members of the Anthroposophical Society on Tuesdays, semi-public lectures on Fridays, in an underground form, which helped unfold a diverse life in Theresienstadt. Since then, much has been written and known about the rich cultural life under the difficult conditions of the Theresienstadt camp; we remember the unforgotten Rabbi Leo Baeck and many others, including Viktor Frankl. Here in Dornach and Arlesheim we also remember the Jewish anthroposophists from Prague, including the painter, Richard

Pollak, who together with his wife, Hilde, worked on the first Goetheanum. Pollak gave over 100 lectures on anthroposophy in Theresienstadt. He died in Auschwitz, his wife in Bergen-Belsen. They tried everything within their limits to the very end to stand up for a different, true image of humanity and the world:

> *Speak, you too,*
> *speak as the last one,*
> *have your say.*
> *Speak ...*
> *But do not separate the 'no' from the 'yes'.* [Celan [76]]

This is what I think the Pollaks did in their own way—a Maria Darmstädter did this in Gurs, Drancy and Auschwitz camps[77], as did many other anthroposophists, which can be verified.

I hope I have also managed in my attempt this evening not to divide the 'no' from the 'yes'. We need to expand and intensify our knowledge of history and our reflections; I see this as an ongoing task, the accomplishment of which can and will help us to make the right decisions in the present and in the future. We too need to 'speak' about the past, the present and the future, making our contributions, in an honest and upright manner. In this sense, this evening's attempt on 9 November was hopefully also a contribution to the *Signature of the Present,* going way beyond aggressive attacks and defamation. Thank you for your attention both here in this room and elsewhere.

[1] ANKE HILBRENNER und CHARLOTTE JAHNZ: *Am 9. November. Innenansichten eines Jahrhunderts. 1918 / 1923 / 1938 / 1969 / 1974 / 1989.* Köln 2020. [9 November. Internal Views of a Century. 1918 / 1923 / 1938 / 1969 / 1974 / 1989.] (German only.)

[2] PAUL CELAN: in *Glottal Stop. 101 Poems by Paul Celan,* translated by Nikolai Popov & Heather McHugh, Wesleyan University Press, 2011.

[3] ANDRÉ SEBASTIANI: *Anthroposophie. Eine kurze Kritik.* Aschaffenburg 2019, S. 164 ff. [Anthroposophy. A short critique.] (German only.)

[4] Quoted UWE WERNER: *Rudolf Steiner zu Individuum und Rasse. Sein Engagement gegen Rassismus und Nationalismus.* Dornach 2011, S. 38. [Rudolf Steiner on the Individual and Race. His Engagement Against Racism and Nationalism.] (German only); Translator's Note: A summary by Uwe Werner: *Anthroposophy in the Time of Nazi Germany* available https://waldorfanswers.org/AnthroposophyDuringNaziTimes.htm (accessed June 4, 2021)

5 HELMUT ZANDER: *Rudolf Steiner. Die Biografie.* München 2011, S. 70. [Rudolf Steiner. A Biography.] (German only.)
6 PETER STAUDENMEIER: 'Der Deutsche Geist am Scheideweg: Anthropo-sophen in Auseinandersetzung mit Völkischer Bewegung und National-sozialismus'. In: UWE PUSCHNER und CLEMENS VOLLNHALS: *Die völkisch-religiöse Bewegung im Nationalsozialismus: eine Beziehungs- und Konfliktgeschichte.* Göttingen 2012, S. 480 ff. [The German Spirit at the Crossroads: Anthroposophists in Conflict with the 'Völkisch' Movement and National Socialism. In: UWE PUSCHNER and CLEMENS VOLLNHALS: The 'Völkisch'-religious Movement in National Socialism: A History of Relations and Conflict.]
7 Refer to the sketch in PETER SELG: *Rudolf Steiner, der Rassismus Vorwurf und die Anthroposophie. Gesellschaft und Medizin im totalitären Zeitalter,* S. 111 ff.
[Rudolf Steiner, The Accusation of Racism and Anthroposophy. Society and Medicine in the Totalitarian Age] (German only. A comprehensive publication is in preparation.)] PETER SELG and MATTHIAS MOCHNER: 'Anthroposophische Medizin, Heilpädagogik und Pharmazie in der Zeit des Nationalsozialismus (1933–1945)'. [Anthroposophical Medicine, Curative Education and Pharmacy under National Socialism (1933-1945).]
8 RUDOLF STEINER: *The Philosophy of Freedom,* 'Individuality and Genus', Ch. 14, Rudolf Steiner Press, 2011. (GA 4) (aka. *The Philosophy of Spiritual Activity,* Anthroposophic Press, 1986.)
9 RUDOLF STEINER: Lucifer—Gnosis. Grundlegende Aufsätze zur Anthroposophie und Berichte aus den Zeitschriften 'Luzifer' und 'Lucifer-Gnosis' 1903–1908. GA 34. Dornach 21987, S. 504.
10 RUDOLF STEINER: *The Mission of the Folk Souls* (aka, *The Mission of the Individual Folk Souls*). (GA 121) Quoted from a Preface written by Rudolf Steiner to members on 8 February 1918.
11 WENZEL MICHAEL GÖTTE: 'Das Verdämmern der Rassen—Rudolf Steiners Individualismus'. In: *Geistige Individualität und Gattungswesen. Anthroposophie in der Diskussion um das Rassenverständnis.* Sonderheft Mitteilungen aus der Anthroposophischen Arbeit in Deutschland. Sommer 1995, S. 27. [The Dawning of the Races—Rudolf Steiner's Individualism'. In: Spiritual Individuality and Human Biology in a discussion about the understanding of race.] (German only.)
12 PETER SELG: *Rudolf Steiner, der Rassismus-Vorwurf und die Anthroposophie. Gesellschaft und Medizin im totalitären Zeitalter.* Arlesheim 2020. [Rudolf Steiner, The Accusation of Racism and Anthroposophy. Society and Medicine in the Totalitarian Age.] (German only, English in preparation.)
13 PETER SELG: *Rudolf Steiner: Life and Work. In:* 'Vol. 4. The Years of World War I (1914–1918)'. SteinerBooks, Incorporated, 2016.
14 RUDOLF STEINER: 'The Spirits of Light and the Spirits of Darkness', in: *Fall of the Spirits of Darkness,* Lecture 12. 26 October 2017. (GA177) Rudolf Steiner Press, 2008.

15. RUDOLF STEINER: *The New Spirituality and the Christ Experience of the Twentieth Century*, Lecture 7. (GA 200) Dornach, 31 October 1920.
16. RUDOLF STEINER: *The Book of Revelation: And the Work of the Priest* (aka: *The Apocalypse*) Lecture 14, 18 September, 1924. (CW346), Translation Johanna Collis. Rudolf Steiner Press, 1999.
17. PETER SELG: *Rudolf Steiner, der Rassismus-Vorwurf und die Anthroposophie. Gesellschaft und Medizin im totalitären Zeitalter*, S.69 ff. Arlesheim 2020. [Rudolf Steiner, The Accusation of Racism and Anthroposophy. Society and Medicine in the Totalitarian Age.] (German only, English in preparation.)
18. RUDOLF STEINER: 'Collected Essays on Culture and Current Events 1887-1901', in *Early Written Literary Work*, Volume 31, of *Rudolf Steiner's Complete Works*.
19. Emil Felber to Rudolf Steiner, 21.9.1897. Rudolf Steiner Archiv, Dornach. Quote freely translated by C. Howard.
20. RUDOLF STEINER: 'Collected Essays on Culture and Current Events 1887-1901', in *Collected Essays, Early Written Literary Work*, Volume 31 of *Rudolf Steiner's Complete Works*. (GA31)
21. RUDOLF STEINER: 'Biographies and Biographical Sketches 1894-1905', in *Collected Essays, Early Written Literary Work*, Volume 33 of *Rudolf Steiner's Complete Works*. (GA33)
22. RUDOLF STEINER: 'Collected Essays on Culture and Current Events 1887-1901', in *Collected Essays, Early Written Literary Work*, Volume 31 of *Rudolf Steiner's Complete Works*. (GA31)
23. PETER SELG: *Rudolf Steiner, der Rassismus-Vorwurf und die Anthroposophie. Gesellschaft und Medizin im totalitären Zeitalter*, S. 46 ff. Arlesheim 2020. [Rudolf Steiner, The Accusation of Racism and Anthroposophy. Society and Medicine in the Totalitarian Age.] (German only, English in preparation.)
24. RUDOLF STEINER: 'Collected Essays on Culture and Current Events 1887-1901', in *Collected Essays, Early Written Literary Work*, Volume 31 of *Rudolf Steiner's Complete Works*. (GA31) (*Gesammelte Aufsätze zur Kultur- und Zeitgeschichte 1887–1901*. GA 31, S. 378 f.—German) Freely translated by C. Howard.
25. Ibid., S. 199. (German.) Freely translated by C. Howard.
26. RUDOLF STEINER: *Building Stones for an Understanding of the Mystery of Golgotha*, Lecture 1, 27 March 1917. *(GA 175)*
27. RUDOLF STEINER: *The Mysteries of Light, of Space and of the Earth*, Lecture 3, 'Historical Occurrences of the Last Century'. 14 December, 1919. (GA 194)
28. PETER SELG: *Der Untergang des Abendlands? Rudolf Steiners Auseinandersetzung mit Oswald Spengler.* Dornach und Arlesheim 2020. [Decline of the Occident? Rudolf Steiner's debate with Oswald Spengler.] (German only.)
29. ALICE WENGRAF: 'Rudolf Steiner als Künstler'. In: Österreichische Blätter für freies Geistesleben, Nr. 4, 1925, S. 19. (German only.)

30 LORENZO RAVAGLI: *Unter Hammer und Hakenkreuz. Der völkisch-nationalsozialistische Kampf gegen die Anthroposophie.* Stuttgart 2004. S.177 [Under hammer and swastika. The peoples' nationalist struggle against Anthroposophy.] (German only.)

31 Ibid.

32 ADOLF HITLER: 'Staatsmänner oder Nationalverbrecher?' In: *Völkischer Beobachter*, 15.3.1921. Wiederabdruck in: Sämtliche Aufzeichnungen 1905– 1924. München 1980, S. 350. [Statesmen or National Criminals? In: *Völkischer Beobachter*, 15.3.1921.] (German only.)

33 In: Archivmagazin. Beiträge aus dem Rudolf Steiner Archiv. Nr. 8. Dezember 2018, S. 200. Review of the German article is available in English online at: http://www.nna-news.org/news/article/?tx_ttnews%5Btt_news%5D=2745&cHash=fc9630f2d6ee29e6545b9123774db227

34 Quoted from LORENZO RAVAGLI: Unter Hammer und Hakenkreuz. Der völkisch-nationalsozialistische Kampf gegen die Anthroposophie, S. 124. [Under hammer and swastika. The peoples' nationalist struggle against Anthroposophy] (German only.)

35 ANNA SAMWEBER: Quoted from her book: *Aus meinem Leben* [From My Life]. Basel 1981, p 44. (German) A book by Jacob Streit, recorded Anna Samweber's *Memories of Rudolf Steiner*. Rudolf Steiner Press, 1991.

36 According to Karl Lang, Steiner also said: 'If this [national socialist] society prevails, it will bring great devastation to Central Europe.' (*Lebensbegegnungen*. Benefeld, 1972, S. 67.)

37 RUDOLF STEINER: *The Christmas Conference,* 1 January, 1924, Dornach. In the evening, Part II: 'The Procedings of the Conference.' (GA 260)

38 RUDOLF STEINER: 'From Nature to Sub-Nature', in *Anthroposophical Leading Thoughts,* March 1925. (GA 26)

39 Ita Wegman to Ernst Lehrs, 30.1.1931. Ita Wegman Archive, Arlesheim.
Refer to PETER SELG: Rudolf Steiner, der Rassismus-Vorwurf und die Anthroposophie. *Gesellschaft und Medizin im totalitären Zeitalter.* Arlesheim 2020, S. 148 f. [Rudolf Steiner, The accusation of racism and Anthroposophical Society and Medicine in the Totalitarian Age.] (German only.)

40 Ita Wegman to Emmy Giesler, 17.3.1933. Ita Wegman Archive, Arlesheim. Translated by C. Howard.

41 Ita Wegman to Fried Geuter and Michael Wilson. 24.3.1933. Ita Wegman Archive, Arlesheim. Translated by C. Howard.

42 Ita Wegman to Daniel Nicol Dunlop, 17.4.1933. Ita Wegman Archive, Arlesheim. Freely translated by C. Howard.

43 WENZEL MICHAEL GÖTTE: Erfahrungen mit Schulautonomie. Das Beispiel der Freien Waldorfschulen. Stuttgart 2006, S. 564. [Experiences with school autonomy. The example of the Waldorf schools.] (German only.)

44 Quoted from UWE WERNER: *Anthroposophists in the Time of National Socialism in Germany (1933–1945)*. An excerpt, with this quote and a citation to the Prohibition of the Anthroposophical Society in Germany, 1, Nov. 1935 is available in English at: https://southerncrossreview.org/82/werner-nazizeit.html (accessed 12, June 2021)

45 Quoted from WENZEL MICHAEL GÖTTE: `Das Verdämmern der Rassen—Rudolf Steiners Individualismus.` [The Dawning of the Races—Rudolf Steiner's Individualism.] S. 16 (German only.)

46 Quoted from UWE WERNER: *Anthroposophen in der Zeit des National-sozialismus (1933–1945)*, S. 383. Translation of quote: C. Howard. [Anthroposophists in the Time of National Socialism.]

47 Ibid.

48 ALFRED BAEUMLER: From 'Report on Waldorf Schools' and 'Report on Rudolf Steiner and Philosophy' by Alfred Bauemler. Original German publication, UWE WERNER: *Anthroposophen in der Zeit des Nationalsozialismus (1933–1945)*. Quote in English found UWE WERNER: *Anthroposophists in the Time of National Socialism in Germany (1933–1945)*, https://southerncrossreview.org/82/werner-nazizeit.html (accessed 12, June 2021)

49 Jakob Wilhelm Hauer to L. Stengel von Rutkowski, 2.9.1940. (P. 301 of Werner's original German publication, UWE WERNER: *Anthroposophen in der Zeit des Nationalsozialismus (1933–1945).*)

50 Die Anthroposophie und ihre Zweckverbände. Bericht unter Verwendung von Ergebnissen der Aktion gegen Geheimlehren und sogenannte Geheimwissenschaften vom 5. Juni 1941. Reichssicherheitshauptamt (RSHA) Berlin 1941. Faksimile in ARFST WAGNER (HG.): Dokumente und Briefe zur Geschichte der Anthroposophischen Bewegung und Gesellschaft in der Zeit des Nationalsozialismus. 5. Band: Nachträge, Briefe, Aktuelles. Rendsburg 1993, S. 56. (German only.) [Translation: Anthroposophy and its Associated Institutions. Report applying evidence from the Operation against Secret Teachings and so-called Esoteric Sciences of 5, June 1941. RSHA, ca. October 1941. Facsimile in ARFST WAGNER (Ed.): Documents and Letters on the History of the Anthroposophical Movement and Society in the Time of National Socialism. 5th volume: Supplements, Letters, News. Rendsburg 1993, p. 56.]

51 Reinhard Heydrich to Richard Walther Darré, 18.10.1941. In: UWE WERNER: *Anthroposophen in der Zeit des Nationalsozialismus (1933–1945)*, p. 327. [Anthroposophists in the Time of National Socialism in Germany (1933–1945).]

52 UWE WERNER: *Anthroposophen in der Zeit des Nationalsozialismus (1933–1945)*. [Anthroposophists in the Time of National Socialism in Germany (1933–1945).]

53 PETER SELG: *Erzwungene Schließung. Die Ansprachen der Stuttgarter Lehrer zum Ende der Waldorfschule im deutschen Faschismus (1938)*, pp. 136 and 258. [Forced closure. The speeches of the Stuttgart teachers on the end of the Waldorf schools under German fascism.] (German only.)

54 Affidavit. Landsberg prison. Copy in the archives at the Goetheanum. E.15.0002.020, in German.

55 PETER SELG: '"Dass die Keimkraft der Idee durch ihre Existenz gefährdet wird …" Anpassung und Widerstand. Die Waldorfschulen im Nationalsozialismus'. In: *Erzwungene Schließung. Die Ansprachen der Stuttgarter Lehrer zum Ende der Waldorfschule im deutschen Faschismus (1938)*, P. 124 ff.
['Adaptation and Resistance. The Waldorf Schools under National Socialism'. In: Forced Closure. The Speeches of Stuttgart Teachers on the End of the Waldorf School under German Fascism (1938).] (German only.)

56 PRIESTMAN, KAREN, *Illusion of Coexistence: The Waldorf Schools in the Third Reich, 1933–1941* (2009). Theses and Dissertations (Comprehensive). 1080. https://scholars.wlu.ca/etd/1080

57 PETER SELG: '"Dass die Keimkraft der Idee durch ihre Existenz gefährdet wird …" Anpassung und Widerstand. Die Waldorfschulen im Nationalsozialismus'. In: *Erzwungene Schließung. Die Ansprachen der Stuttgarter Lehrer zum Ende der Waldorfschule im deutschen Faschismus (1938)*, S. 161 ff.
['Adaptation and Resistance. The Waldorf Schools under National Socialism'. In: Forced Closure. The Speeches of Stuttgart Teachers on the End of the Waldorf School under German Fascism (1938).] (German only)

58 ALEIDA ASSMANN: *Das neue Unbehagen an der Erinnerungskultur. Eine Intervention.* München 2016, S. 92. [ALEIDA ASSMANN: The New Discomfort with Memory Culture. An Intervention.] (German only.)

59 Ita Wegman Archive, Arlesheim.

60 Ita Wegman to Franz Löffler, 16.1.1935. Ita Wegman Archive, Arlesheim.

61 PARACELSUS: *Sämtliche Werke*. Band 6. Ed. WILL-ERICH PEUCKERT. Darmstadt 1965, S. 55. [Complete Works, Volume 6.]

62 See, PETER SELG: Erzwungene Schließung, S. 137 ff. and S. 260 ff. [Forced Closure.] (German only.)

63 PETER BIERL: Wurzelrassen, Erzengel und Volksgeister. Die Anthroposophie Rudolf Steiners und die Waldorfpädagogik. Hamburg 2005, 18. [Root Races, Archangels and Folk Spirits. The Anthroposophy of Rudolf Steiner and Waldorf Education.] (German only.)

64 LORENZO RAVAGLI: 'Polemischer Diskurs: die Anthroposophie und ihre Kritiker'. In: PETER HEUSSER UND JOHANNES WEINZIRL (HG.): *Rudolf Steiner. Seine Bedeutung für Wissenschaft und Leben heute*. Stuttgart 2011, S. 345 f. [L. Ravagli, 'Polemic Discourse: Anthroposophy and its Critics'. In: PETER HEUSSER AND JOHANNES WEINZIRL (eds.): Rudolf Steiner. His Significance for Science and Life Today.] (German only.)

65 RUDOLF STEINER: Gesammelte Aufsätze zur Kultur- und Zeitgeschichte 1887–1901. (GA 31), p. 382. [In Complete Works: Collected Essays on Culture and Current Events 1887-1901.] (Translation: C. Howard.)

66 LORENZO RAVAGLI: 'Polemischer Diskurs: die Anthroposophie und ihre Kritiker'. In: PETER HEUSSER UND JOHANNES WEINZIRL (HG.): *Rudolf Steiner. Seine Bedeutung für Wissenschaft und Leben heute.* Stuttgart 2011, S. 332–352. [L. Ravagli, 'Polemic Discourse: Anthroposophy and its Critics'. In: PETER HEUSSER AND JOHANNES WEINZIRL (eds.): Rudolf Steiner. His Significance for Science and Life Today.] (German only.)

67 RAINER MAUSFELD: *Warum schweigen die Lämmer? Wie Elitende Demo kratie und Neoliberalismus unsere Gesellschaft und unsere Lebens grundlagen zerstören.* Book pubished, Frankfurt 2019, p. 35. [Why the lambs remain silent? How elite democracy and neoliberalism is destroying our society and the foundations of our existence]. **Translator's note**: An extended version of a talk presented at the Christian-Albrechts-Universität Kiel, 22. Juni 2015 by Prof. Mausfeld has been translated into English by Dr Daniel Wollschläger giving English readers valuable insight into Rainer Mausfeld's work, entitled: 'Why do the lambs remain silent? On democracy, psychology, and the ruling elite's methods for managing public opinion as well as public indignation.' Available online, accessed 16.6.2021
https://cognitive-liberty.online/wp-content/uploads/Mausfeld_Why-do-the-lambs-remain-silent_20151.pdf

68 Ibid. p. 160.

69 In: TED A. VAN BAARDA (HG.): Anthroposophie und die Rassismus Vorwürfe. Der Bericht der Niederländischen Untersuchungskommission 'Anthroposophie und die Frage der Rassen'. Frankfurt a. M. 2009, 131. [Anthroposophy and the accusations of racism. The Report of the Dutch Commission of Inquiry 'Anthroposophy and the Question of Race'.]

70 RUDOLF STEINER: 'The Year of Destiny 1923 in the History of the Anthroposophical Society. From the Burning of the Goetheanum to the Christmas Conference', in *Complete Works.* (GA 259) Not yet available in English. Quote translated by C. Howard.

71 LOUIS M. I. WERBECK: Die christlichen Gegner Rudolf Steiners und der Anthroposophie durch sich selbst widerlegt. Stuttgart 1924, S. 110. [Christian Opponents of Rudolf Steiner and Anthroposophy Refuting Themselves.] (German only.)

72 RUDOLF STEINER: *Ideas for a New Europe: Crisis and Opportunity for the West,* Rudolf Steiner Press, 1992. (GA 196) and as, *The History and Actuality of Imperialism,* 3 Lectures, 20/21/22 February 1920 (GA 196).

73 RAINER MAUSFELD: Warum schweigen die Lämmer? Wie Elitendemokratie und Neoliberalismus unsere Gesellschaft und unsere Lebensgrundlagen zerstören, S. 128. [Why the lambs remain silent? How elite democracy and neoliberalism is destroying our society and the foundations of our existence.] (Book in German.)

74 Cf. PETER SELG: Nach Auschwitz. Auseinandersetzungen um die *Zukunft der Medizin.* Arlesheim 2020, S. 63 ff. [After Auschwitz. Debates about the Future of Medicine.]

[75] PAUL CELAN: *Gesammelte Werke in fünf Bänden* [*Collected Works* in 5 Volumes, in German.]

[76] PAUL CELAN: *Speak, You Too.* Translator unknown. Found online, accessed 16 June 2021: https://somepoems.livejournal.com/9176.html

[77] Cf. PETER SELG: *Maria Krehbiel-Darmstädter. Von Gurs nach Auschwitz. Der innere Weg.* Arlesheim 2010 (in German).

Constanza Kaliks / Paula Edelstein

Can what is foreign be understood?

In conversation, Constanza Kaliks and Paula Edelstein explore how interculturality is emerging as a path forward for the twenty-first century and the particular challenges facing education.

Constanza Kaliks: The pandemic has not only presented itself as a problem, but once again raises the possibility, indeed the necessity for examining problems that have been in the world for decades. One of these major challenges is the increase in conditions of extreme material deprivation and vulnerability. Not all people come to earth in situations that give them the opportunities that a humane environment would. This fact has been exacerbated by the pandemic. An example of this is described in the UNESCO report of August 2020: in addition to the millions already suffering from hunger, the pandemic has added 150 million more young people and children because they have not had their daily meals due to school closures. The report states: 'It is a universal crisis and for some children the impact will be felt for the rest of their lives.'[1]

How can we deal with the knowledge of this situation? It is given to us and we can inform ourselves about the situation that has arisen in recent months. The possibility of knowing also brings with it responsibility. The pandemic has further highlighted our mutual responsibility and our common tasks have become clearer. We are thereby called upon to take this on together, each person within the environment in which he or she can be most effective. The pandemic has shown our interdependence and reciprocity, which is actually what being human is. It shows that we are really dependent on each other. We are actually involved in this reciprocity just by being in the world. How can we take on this task of destiny? The question of the 'otherness' of the other person presents itself to us in a radical way, because it is actually a fact of daily life. Indeed, how can we really understand the 'other'? Are we able to embrace the challenge of mutuality, if the other person is a stranger to us?

We Are in this Reciprocity Just by Being in the World

Can we develop insights that embrace life and which become an orientation for responsible action—in which, not I with my own frame of mind, with my own perspective on things and in which the world becomes my yardstick, but rather, in which I and my views stand in relationship to others, I and my destiny are interconnected to the destinies of society and nature?

In this conversation, we want to focus on three aspects of this huge topic in the field of education. First, the aspect of our plural belonging, which is simultaneously and complexly interwoven. Then the aspect of mutuality, which is always concrete, always has a name and a place, because we are beings in a situation. And finally, the aspect of the uniqueness of each human being, each individual as an absolute unique reality.

Plural Belonging

Every human being enters the world into an environment that brings with it many natural affiliations, affiliations that unfold and interweave. We are conscious of many of these affiliations or things that bind us together—family, background, language, the immediate environment in which we experience the world—but there are many others which remain unconscious which are often just as formative and present. The experience of uprootedness, homelessness and alienation affects many people around the world today. What happens when these natural affiliations are not there, do not occur?

Paula, you have had many experiences in your educational work with people and schools in the suburbs of Buenos Aires and have examined them in your research. What has become apparent to you?

Paula Edelstein: We discovered the issue of interculturality when we asked ourselves 20 years ago what the reasons were for the pupils' inability to learn at school. They were children living in the peripheral neighbourhoods of Buenos Aires where there is a lot of poverty. Children who could not learn at school, who were therefore 'school failures', were children who were awake and present in their own streets, in their own environments. There they were alive, they were interested, and they asked questions. At school, however, they displayed completely different attitudes. They were aggressive and refused to participate, or were very reserved, quiet to silent, or did not participate at all. The teachers felt powerless and frustrated. We then wondered if the school was not asking the right questions. Does the school really know who their pupils are? Does the school know the pupils at all? It became evident that there was a huge chasm between what the school was proposing and the reality in which the children lived. Given this chasm, it was impossible to address the children. This chasm had been created over decades, as a result of persistent discrimination: against the indigenous people and against the internal migrants from the provinces or from neighbouring countries. The distance was the result of the constant denial of potentiality, of making the qualities of the other invisible. The non-recognition of self-identity is a fundamental aggression, because it is a violation of one's own being; and the school did not know the pupils with whom it worked.

With this panorama, we posed a new hypothesis: we thought that this failure at school probably had nothing to do with the qualities or potentialities of the pupils, but with cultural alienation and misjudgement, which did not free them to express themselves and be who they are. We understood that this is how dissimulation is born, as a process of defence, and we saw that this population hides what it is because it is considered shameful by the system: their roots, their origins, the colour of their skin, the places where they live, everything is seen as shameful. That's why families hide their true identity, which they develop in their own protected spaces, away from the outside world, because there they are exposed to discrimination and violence. And they teach their children to try and be like the 'others' in the world in order to belong, to fit in.

The attempt to become similar in order to belong is hopeless, because it is not possible to hide one's own feelings of belonging without paying a huge price. Either it does not succeed and the others realize that we are not what we appear to be and deny and discriminate against us all over again. Or, it succeeds and we appear to be so similar that we forget who we are and become lost to ourselves.

Every Human Being is an Absolute Unique Reality

In order to reverse the failure at school, it was necessary to observe and to listen, with the desire to know and to allow their real identities and true affiliations into the school. In order for education to take place, one has to be able to reveal oneself, one has to be able to make oneself known, and for this to happen one has to be able to unveil oneself.

At first, trust had to be built and discrimination turned around. It was we, as teachers, who had to bring images that the children and their families talked about at home into the classroom: the cotton fields, the mountains that their grandparents remembered from their provinces, the carts loaded with cardboard that they used to collect and move around the city every day, the little places between the streets where they set up a place to play ball. The streets of the neighbourhoods where the children had played, grown up in and learnt in had to come into the classrooms so that we could learn from them and observe through them.[2]

Mutuality, Encounter, Location

Constanza Kaliks: Rudolf Steiner's educational impulse is based on the search for knowledge of the child, in recognition of his or her 'becoming' with and from the multiple, interwoven affiliations anchored in their physicality, in his or her life of soul and in his or her distinctive uniqueness as a spiritual being.

The place where I learned to see the world, to listen to it, where I learnt to learn, is always very concrete and real. Who was I connected and close to, who told me about the world, who enabled me to see it? Culture is anchored in what is perceived. If we want to give attention to the intercultural, we have to observe the place, the location and the real life of the other.

Rudolf Steiner addresses this concrete aspect of the child's environment:

> For the point is not to think out some way in which a number of children may be educated quite apart from the world, according to one's own intellectual, abstract ideas, but rather to discover how children may be helped to grow into true human beings within the social milieu which is their environment. One must muster one's strength and not take children away from the social milieu in which they are living. It is essential to have this courage. It is something which is connected with the world significance of education.[3]

How do we develop an education that takes this reality of connection as the basis for learning: this being-in-the-world? This relationship to the other and to the world is the foundational substance from which to draw. The deep longing to know the other and the world is based on this. Without relationship, there is no reason to discover or to learn. The intercultural is the intermediate sphere, the space that will always emerge when this relationship is enacted. It is not a place that arises when simply two opposite poles face each other, but is formed out of the reality of the accomplished relationship between these two or more.[4] How can life be practised and shaped in this in-between sphere?

Paula Edelstein: We have the impression that the classroom is a space that can facilitate this in-between sphere as an experience. Every time we were able to connect with genuine interest to the place we were working on, we discovered new expressions, stories, narratives, new affiliations and worlds that could be brought into the classroom. The people who live in these marginalized areas come from very far away and bring their places with them—the high plateaus of the Andes or the forests along the great rivers. These could now enter the classroom, along with their ancestors and with millennia of knowledge. The families create this anew when they come to the big city. When we managed to overcome their hiding away and built a bridge, we could speak of many things. We spoke of the impenetrable forests, in the memory of their grandparents, and also of the pain of knowing they had been cut down. We talked about the rivers and the fishermen, about the pollution of the rivers that kills the fish. We talked about their need to leave their areas in order to survive, and the strategies they are using here to do this with maximum creativity and minimum resources; for example, the community kitchen where everyone gets something to eat, or the collection of waste paper to sell. As they talked about all this, the classroom filled with content. The concept of work and the environment changed their meaning; they became deeper, more complex and more real. It was possible to ponder origins and futures, to think of other forms and open new horizons.

> In order for education to take place, one has to be able to reveal oneself, one has to be able to make oneself known, and for this to happen one has to unveil oneself.

Through the voices of the children, the stories of their families, the deep experiences of the people of our continent arrived at the school. It is an experience that grows and grows by observing and accompanying. It was not only the experiences of the children, but it was a whole system of knowledge that entered the school.

Learning in this way, a forest cannot be confused with an oxygen factory, because there are the living memories of that cleared forest, of the pain brought by its loss. If the engineers and geologists of today had learnt about the forests with love and respect as children, would they continue open-cast mining as destructively as they do today—or would they find alternative ways?

When we facilitate intercultural dialogue, we can foster understanding and respect for the territories of the world, broaden horizons and open borders. By listening and speaking out, our identities are strengthened in such a way that the desire to know more grows and grows in respectful ways. As they begin to narrate, the children also begin to ask questions, become interested in other things and to want to know about the others.

The Uniqueness of the Other

Constanza Kaliks: Finally, we would like to address the task of understanding the other from the reality of the other. Here we face an immense challenge: can I make an inner active sacrifice so that the being of the other can express itself? Another form of reference is needed, a dialogue, a speaking that listens: an intimate attentive listening. An anthropology of dialogue is needed to learn to understand the other from out of their circumstances. We know that the norms of one culture are not enough to understand the norms and reality of another culture.

Steiner's pedagogical approach is based on an understanding of the human being that respects the other, as a riddle, as an ongoing mystery as well—an education that not only makes itself accessible to an existing reality, but one that affirms and supports every young person that enters the world in their becoming, in everything that does not yet exist in them.

This is the space in which the uniqueness of every person can be experienced, in that each one is irreplaceable. It is not about the human being in general, but about the concrete, real, unique and singular human being in whom the dignity of being human is expressed. This is also reflected in the call for anti-racism after the recent murders in North America: 'Say his name—say his name!' It is not about anyone or a person in general, but about this person, this unique one, the irreplaceable essence of this human being. This one is not an example of a human being, but she/he is a unique human being with her/his unique way of being in the world. In a story, the Mozambican writer Mia Couto wrote:

> *When asked about his race, he answered:*
> *My race is me, João Passarinheiro.*
> *When asked to define himself, he added:*

My race is I, myself.
Each one is an individual human being.
Each human being is a race, Mr. Policeman.[5]

In his book, *The Philosophy of Freedom*, Rudolf Steiner affirms the absolute uniqueness of each one: from the point of view of his unique identity, his spiritual constitution—each human being is his own unique species.[6]

Paula Edelstein: In the world today, in which so much is changing, we have asked ourselves what position the school wants to take and can take. Today, we need to understand education as a period of working and building something together, in which each one of us, as individuals, unfolds our being in the world. We need to understand school as the place of encounter that makes this shared work possible, something which is our own and something which is with others. It is essential to create spaces in the classroom for the rich and varied expression of the images and territories that each of us carries with us. The different approaches to learning need to enter into dialogue. The power of the circle is very crucial for this dialogue. In the Waldorf school, we are invited into this circle; we can think of what happens in the classroom as a circle event—not just as a game, but as a form of coming together. In this form, words flow, horizons and points of view mingle. The cognition that emerges is diverse, parity-based and plural. Asymmetries are broken down, not only through the voices in the classroom, but also through the affiliations that each one brings with them. It allows a knowing of other ways of knowing, born from other stories, and thereby also learning new things about myself, expanding my own affiliations and developing with the being of others, together.

It was very enriching to hear these voices that had been silenced or hidden for so long. It was not only enriching because I discovered the worlds of these children. I also discovered the very complex systems of knowledge that the indigenous people, for example, carry within them. It was enriching because I discovered aspects of myself and my own history through this dialogue. I expanded my own sense of belonging; I understood my own social being better. When I brought these voices into other classrooms—I could teach astronomy, for example, by putting the Andean calendar next to the Western one, and as a result, the whole content expanded. The young people I interacted with recognized more elements within themselves and could therefore express themselves. So the desire to want to get to know the others grew.

Constanza Kaliks: The *other*: are we learning to do more than tolerate—are we learning to want it? Are we learning to become aware of our mutuality as an element that is part of us? And this not only in our social becoming, but also in individual becoming and in our relationship to nature, to the cosmos, as a generator of climatic conditions, for the dignity of the human being and the world? Are we learning to maintain our bonds and recognize them as the basis of our common and individual being in the world? So that, as the Brazilian educator Paulo Freire says, the vocation of 'becoming' 'is not a privilege of the elite, but the birth right of all human beings?'[7]

The space at the school seems to provide an outstanding learning space for great human tasks. Having come from our different life circumstances, to a shared and common 'now', we are learning to contribute to these great tasks: in respect and in recognition that everyone is unique and irreplaceable, by recognizing our reciprocity as a constitutive part of reality and by understanding the others from within themselves.

[1] UNICEF: Education and COVID-19, accessed on 18 June 2021, https://data.unicef.org/topic/education/covid-19/

[2] María del Carmen Maimone und Paula Edelstein, *Didáctica e Identidades Culturales: acerca de la dignidad en el proceso educativo.* La Crujía, Stella, Buenos Aires 2004. [Didactics and cultural identities: about dignity in the educational process.] (Spanish only.)

[3] Rudolf Steiner: *Human Values in Education,* Lecture 6, Arnhem, 22 July 1924. (GA 310).

[4] Cf. Waldenfels, apud Georg Stenger, Ort/s, Ortungen, Orientierungen. In: Murat Ates et al. (eds.), Orte des Denkens - [Places of Thinking]. Karl Alber, Freiburg and Munich 2016, p. 28. (In German.)

[5] Mia Couto, 3. Ed., 'Cada homem é uma raça, senhor polícia.' [Each human being is a race, Mr Policeman.] Editorial Caminho, Lisbon 1994. (In Portuguese.)

[6] Rudolf Steiner, *The Philosophy of Freedom.* (GA 4), Rudolf Steiner Press, 2011. (aka.: *The Philosophy of Spiritual Activity,* Anthroposophic Press, 1986.)

[7] Paulo Freire, Pedagogy of the Oppressed. Available online, accessed 19 June 21: https://ia600204.us.archive.org/24/items/PedagogyOfTheOppressed-English-PauloFriere/oppressed.pdf

Matthias Rang

Are we making a religion out of science?

At a youth conference in Freiburg two years ago, which explored the question of how we want to shape our future, the young people formulated a clear consensus: there will only be a humane and positive future if society lives much more through and with art and culture. A positive shaping of the future needs to be a cultural shaping. As a natural scientist, I then asked what role science will play in this future. There were two common opinions on this from the young people: the first was that science is very important and should be viewed in a positive light. It allows us to understand things and processes, and at the same time provides us with technology that can be used to shape the future. The second opinion was rather critical of science because it has given us technology whose use has led to considerable problems, both social and environmental. We can also observe how we are increasingly becoming dependent on technology and how our thinking is also becoming increasingly technical.

The two opposing opinions were well justified and each was supported with numerous examples. At that time, we concluded: the two opinions that will prevail will be less dependent on science itself and the questions scientists ask themselves, but would, in particular, depend on how we deal with science and its findings, on what significance is allotted to them and what role is assigned to them by society. This will be decisive in determining whether science has a positive effect on the shaping of our future or whether it is more impeding.

Now two years have passed since this conference and we are experiencing a great paralysis in society and we are occupied with one theme, the Corona crisis. Topics in public debate have also shifted considerably. For example, the climate crisis, which reflects the two possible roles for science in our future with its reference both to knowledge through science and at the same time, with its connection to industry and mechanization as components of the problem, has currently largely disappeared from public attention.

Currently, in the midst of the Corona pandemic, it is all the more clear that we are dependent on science as a tool for acquiring knowledge and that it must and will play a very particular role in assessing situations and solving problems. The question I will try to address in the following is: Under which conditions can science be positively effective in this global challenge?

What is the basis for certainty of knowledge in science?

What is the characteristic trait and special feature that defines science? If I try to give a very brief answer to this, I think I can say that in the course of human development in the natural sciences, we have been able to gain a certain kind of certainty of knowledge about nature and the structure of its relationships. I think this certainty of knowledge is rightly claimed. It arises from the fact that we know how a scientific result comes about—quite precisely—and can therefore also state in detail afterwards how the result came about. The aetiology of this result is, to a certain extent, transparent in science, because if this is the case, I will always be able to arrive at a similar outcome in a new experiment.

The crucial point here is that the aetiology of scientific knowledge—the history that leads to this result—namely the conditions and the preconditions for the conclusion arrived at, are already a central part of the conclusion itself. That means: every scientific statement of fact already contains its own aetiology within it.

This can be illustrated with a simple example. If I take a stone and hold it one metre above the ground and drop it, the stone will hit the ground in about half a second. The time can be specified quite exactly. The result can always be reproduced and verified, because the statement itself contains the aetiology of the result, for the word 'if' introduces the description of the condition underlying the result. This results in the certainty of knowledge, the scientific fact. If, on the other hand, I do not comply with these conditions, for example if I take a feather instead of a stone, for which further conditions would have to be taken into account, then the certainty of knowledge is no longer given. The statement can even be false and the process can take a completely different course.

This is therefore a special kind of statement in which the conditions are included in the statement. In the theory of science, this is referred to as a conditional statement or conditional judgement. In this methodology, science is closely related to mathematics, for the structure of mathematical statements follows the same principle, as the following example shows: if I have three quantities, A, B and C, and if A is larger than B and if B is larger than C, then A is also larger than C.

Therein is the certainty of knowledge. It does not say at all whether I have these three quantities A, B and C. The condition mentioned is an assumption, or rather a prerequisite, and then it is important that two further conditions are fulfilled, namely that A is greater than B and B is greater than C. Only then is A greater than C. If, on the other hand, I omit one of the conditions in the mathematical statement, then the statement that A is greater than C is completely stripped of its certainty of knowing.

This brings us to a first, important point that is not sufficiently present in many presentations supposedly based on scientific results, even in many articles about scientific problems. Natural science grants its certainty of knowledge only in the structure of 'if' and 'then', and only if this 'if' and 'then' are not only fulfilled, but are also included in the statement itself.

Limitations to the Certainty of Knowledge in Science

Scientific theories that do not have the structure of a conditional statements, but make causal statements, i.e. depict causal statements or causal judgements, must be very carefully and consciously distinguished. These theories no longer have the descriptive character that corresponds to the empirical scientific statement, naming its conditions, rather, it looks for causes. If we think of the example of the falling stone, then answers to the question are provided by such theories: Why does the stone fall to the ground? The answer to this question depends more on the theory than on the phenomenon itself. In the context of gravitational theory, this question would be answered by the earth's gravitational force—and this regardless of whether a stone or a feather falls from whatever height and under whatever circumstances. Whether or in what form gravity exists or is only a functioning hypothesis is and remains completely open, in principle. There is no certainty of knowledge in this area. Every theory that has been established in science has been overhauled in the course of time or has had to be reformulated in its assumptions and principles, and so it is with all current theories.

Goethe already clearly recognized the difference between the certainty of knowledge of empirical scientific statements and hypothetical theorizing and sought to ban theorizing from the discipline of natural science.[1]

In doing so, he also presented a critique of 'claim of proof' in science, which was later specified and theoretically well-substantiated by the physicist Pierre Duhem. The philosopher of science, Karl Popper, also presented a similar critique of 'claim of proof', stating that theories can never be verified or proven, but can only be falsified, that is, disproved.[2] Theories can be recognized as false, but not as true. Therefore, there is no certainty of knowledge for this area of natural science: either a theory is proven false and is no longer interesting, or it is uncertain whether it is false or true.

This is a second point that is often not sufficiently differentiated in many popular scientific descriptions of the scientific method, and indeed in scientific studies as well. Due to a lack of clarity in this distinction, empirical facts are intermingled with theory and theoretical conclusions, so that the factual situation, recorded descriptively, might be regarded as reliable or certain, but is however compromised in its statement.

Generalizations, Fake News and Erroneous Trends

But let us leave out the realm of theories for further consideration and stay entirely with empirical scientific facts, as they are recorded, for example, in a scientific study, be it on the climate crisis or on Corona. As such, they only retain their certainty of knowledge if I formulate the conditions under which they apply and take cognizance of them at the same time. If I omit the conditions, the result immediately loses scientific significance.

Against this background, it is problematic how scientific results are often communicated and how they are expressed and disseminated in public forums, in journals and in the media. In these areas today, a certain tendency can be observed: a scientific result, which absolutely provides certainty of knowledge, is not only detached from its conditional structure, but beyond that, is also generalized. The result is then claimed for cases that are even contradictory to the original conditions. Therein lie two steps that are not permissible with regard to certainty of knowledge: first of all, I leave out the conditions for the specific result (thus depriving the result of certainty of knowledge), and then secondly, to generalize the statement for all possible cases. Although, scientific certainty is then claimed as a given, but in many cases, scientific certainty, in fact, no longer exists.

This practice in the media is also encouraged by the often accompanying misunderstanding, that a generalized result offers a higher epistemological value and more certainty than a limited and concrete result. But exactly the opposite statement is correct: every generalized result is based on concrete statements and in principle offer less epistemological value, are mostly indeterminate in their scope and applicability, and, in addition, take on the uncertainties of scientific assumptions, hypotheses and theories. Only the charm [sic] of general validity belies these deficits. In addition to the two points already mentioned, there is a third issue here that is not adequately presented in public perception and reporting on scientific topics, and which makes it difficult, in some cases even impossible, for the interested public to form a healthy and independent judgement.

From this fact, it is quite understandable, that some readers, when faced with media reports, have no wish to pursue them, because, as shown above, these draw generalized conclusions from specific knowledge, claiming the same certainty as the empirical statements. Instead of making the aetiology of the scientific statement visible, an exegesis of this statement is sold as fact. In extreme cases, an argument can actually be generated from a statement in the public media—which borders on fake news—that was originally carefully backed up by scientists, and this, not due to the generalization of the content itself—which may be current and legitimate—but due to the claim that this generalization of content has an unconditional and general validity with scientific legitimacy. Under these circumstances, it is no wonder that reader groups, especially people with an academic background (as has been suggested by recent research in the social sciences), do not wish to follow such media coverage, while other reader groups follow such coverage, resulting in an inevitable polarization of society.

It seems crucial to me here that this societal polarization does not even need to be attributed to the divergent assessments of the experts—in some cases these can be largely congruent—but can be caused, at least in part, by the way they are presented with their false claims arising from the presentation.

The polarization of society combined with accusations of fake news—which has become completely out of hand in all directions and groups, and which has been carefully substantiated in a few instances only—is also the result of this condensed type of reporting in the media, in which the scientific statements are in fact implausible. There is nothing wrong with the generalization of empirical statements that are in themselves correct and concrete. However, in their generalized form, they are usually no longer entirely correct, but are to be regarded as an approximation, a plausible assumption. Therefore, it is also not justified that a questioning of a generalization—mostly on the issue of certainty of knowledge that is not given in the generalization, and not even a contradiction to it—is very quickly dismissed as fake news.

Is science a theistic reflex or a substitute for religion?

Thus, we notice how a certain dynamic has entered society, in which scientific communication is not without blame; on the other hand, by contrast, science could have a positive effect and be accompanied by an anticipation of certainty of knowledge. In the current abridged and condensed form of media communication, however, it has become very problematic. How often does a news item, a commentary or a report begin today with the ever valid, all-purpose sentence (a most perfect example of generalization): 'A recent scientific study has shown that ...'? The detailed conditions are almost never mentioned, but the result appears like a revelation of an absolute scientific fact, which then leads to the aforementioned phenomenon of the polarization of the various social belief groups.

When such a result, which originates from a really good scientific study, is presented in this form, it suddenly takes on the character of a timeless and unconditional truth, a general revelation, having a character that we would never claim as scientific fact, but at most would be ascribed as religious teaching. That is why today we have to ask the question whether, through this public mode of scientific referencing, we are not actually making a kind of pseudo-religion out of natural science? In other words, a certainty that was given in the past to society, as religious truth, is being replaced by apparently new and modern kinds of truth, which, however, do not exist in the way they are being claimed to exist. Natural science, as a societal norm, is replacing a space increasingly vacated by religion, as it were. So we are dealing with a social phenomenon in which science and its players are not uninvolved. However, it is not primarily a scientific problem—many scientists are aware of the limitations of their claims—but a societal problem and the correct assignment of the role of science, its results and our perception of them.

The Hamburg painter and anthroposophist Karl Balmer (1891–1958) once described the danger of such a perception of science in a very succinct and concise way. He portrayed the longing for an *unconditional*, that is a certainty of knowledge *without conditions* in science, as a 'theistic reflex'.[3] In other words, a reflex towards God in fact, a God reflex. It is perhaps not surprising that such a reflex becomes particularly visible in situations in which humanity is existentially threatened and therefore seeks security and a firm foothold. In the past, religions and ecclesiastical denominations were able to provide this security; nowadays, science has to serve this need to a certain extent. But science only offers a 'conditional' certainty—being gained by limiting it to concrete conditions—and is precisely not a general or 'unconditional' certainty of knowledge and therefore cannot take on this role in a generalized sense.

I would like to give a brief example, in which one can well observe this radical kind of 'theistic reflex'. This is in the coverage of some authors in the sceptic movement. Naturally, there are wonderful sceptics who live up to their reputation and actually approach everything in the world with a questioning, sceptical attitude. However, there are also some sceptics who use science like a kind of pseudo-religion, by using results that arise from completely different areas, and applying them, for example, to Waldorf education or complementary medicine. This leads them to an accusation that the scientific results contradict the content of Waldorf education or complementary medicine and that these are therefore completely absurd and merely spring from religious beliefs. It is sometimes really painful to read such things as a scientist. For those concerned do not realize that they themselves have fallen prey to precisely the same argumentation, which they seek to apply to Waldorf education or complementary medicine, by taking the results of a particular study, stripping it of the conditional structure of the study, and thereby stripping it of its inherent certainty of knowledge, and then imposing it on something else in an isolated and generalized form. However, the sceptic movement is itself as diverse as the anthroposophical movement (and I do not want to fall to the generalization trap I have just outlined), so I do not want to pass a disqualifying judgement on the sceptic movement as such. But the example shows that this difficulty in dealing with scientific results is of a fundamental nature and goes beyond mainstream media coverage of scientific issues or the problems of the Corona pandemic that so preoccupy us at present. In all these manifestations, I believe, a certain signature is evident, a *spiritual signature of our time*, the resolution of which would be of inestimable societal value.

What can science do for society?
From Generalization to Contextualization

Against this background, how should we deal with science so that it can have a positive effect and contribute to a favourable societal development in the future? Due to the omnipresence of scientific findings in the media and their influence on political decisions, there is much to be said for the reduction of the presence of science in public and societal decisions. But from what I have presented, I hope to be able to show that what we need is more science, and *not* an increase in generalized conclusions in the media, but a more *scientifically correct handling* of concrete findings. This will not only help the public to appreciate the results in the right way, but will also enable the assessment of the limitations and constraints of the scientific conclusion. Such a sober attitude to knowledge and the awareness of the dangers of generalization, is also necessary with regard to the question of the possibilities opened up by scientific enquiry.

For the scientific problem of the generalization of science—that draws its knowledge by focusing on a limited subject matter and specifying a complete set of preconditions—is not only a problem of reporting and public communication, but is also a problem of the applicability of that result in a situation, where perhaps only a single, but critical precondition deviates from the set of conditions under which the result was arrived at. Very often, as in the case of the Corona pandemic, many critical preconditions are neither known during the emergence of knowledge, nor during its attempted application.

This can hardly be seriously disputed from a scientific point of view and is already sufficient as a statement to show the far too flippant use of fake news accusations from all sides. To mitigate these socially divisive tendencies, a perception of science that is less focused on belief and more on scientific methodology, built of descriptive conditional statements, would be of great importance. However, an understanding of science developed in this direction could also change political decision-making. If one takes the character of scientific accuracy seriously, as I have tried to present here, then a scientific result, with certainty of knowledge, is only valid in the setting in which it was obtained. This certainty is lost when applied to a complex situation. Of course, we sometimes have no alternative but to act! But it seems to me that clearer communication of the real situation would lead to less polarization in society, much less opposition, and would enable a freer supportive sharing. There would be a stronger sense of a common purpose. It appears to me that we, as members of society or as people in general, are not trusted enough to face such a situation calmly and positively.

Nevertheless, I am of the opinion that it is desirable and necessary to develop yet another cultural step in this regard: a kind of 'certainty of action'. This would not be a scientific problem, but would only be an achievable developmental step, socially and culturally. A kind of certainty of action, which of course by its very nature and complexity, could only ever remain provisional and imperfect, could only arise after taking into account the autonomy of the sector in which we act; just as with certainty of knowledge, this can also be achieved by specifying the preconditions. When we act in the cultural sphere, it is the inherent laws of culture that apply. If we operate in another social dimension, then it is the intrinsic aspects or characteristics of this social dimension that would have to guide our actions. This culture, as a process of decision-making and action, only hinted at here, could initially be developed by replacing generalizations with recontextualization. Because in current scientific knowledge, there is only a small, direct determining context of the conditions themselves. Broader contexts—in particular with accompanying cultural circumstances, which are of central importance in our society, but may be meaningless for scientific focus—are completely outside the specific context. In reality, action is often inevitably taken in the larger context, in which generalizations prove again and again to be inadequate, unsatisfactory and unfounded within the framework of the broader context. On the other hand, a yet-to-be-developed culture of contextualization or recontextualization would take these cultural dimensions, this broader context, into account beforehand, instead of only recognizing them afterwards in the social-consequences phenomena of this generalizing way of taking action.

Interestingly, there are approaches for doing this, and these approaches are already over two hundred years old, going back to Goethe. Goethe first tried to expand science in this direction, as a tool of cognition, by including broader conditional structures, and he also gave hints as to how a broader context could also change our ways of acting. The search for an expanded concept of science is still comparatively young—one which definitely still expects and also welcomes the fact that we do precise, predefined studies that go into great detail for gaining a certainty of knowledge through defining limitations, but which also always practises the inclusion of the entire context, and never completely loses it. The toolbox for methods is also expanded in this process, for example, by including artistic methods alongside purely quantitative ones. It is precisely these methods that are often applied in the criticism of the goetheanistic concept of an expanded science.

What I see here in its germ phase, and what I have called a yet-to-be-developed recontextualization process, seems to me like an inverted Goetheanistic cognitive process. In the Goetheanistic-expanded process of cognition, an abundance of modified phenomena, from a never-ending sequence of metamorphoses, is interpreted as an inner law. Instead of generalizing the phenomena in practical application, recontextualizing them would modify and transform them into the reality of society, in accordance with their respective conditions, including their cultural and social ones.

Especially in the current situation, we can see that we need an expanded science that builds on what we have been able to do in science thus far, and beyond this, tries to do even more of what we have not yet been able to do and what still needs to be done. Still, I think that even this expanded science can only become positively effective in situations like the present one, if we also try to further develop our ways of acting in accordance with an inverted procedure of contextualization learnt from science thus far.

[1] Cf. for example Goethe, J. W.: 'The Experiment as Mediator of Object and Subject.' In: *The Nature Institute, In Context,* #24, Fall 2010, p. 19. Translation by Craig Holdrege.
https://static1.squarespace.com/static/5d41f82684370e0001f5df35/t/5f3f-153baa34794fc9b090d3/1597969723327/Goethe+The+Experiment+as+Mediator.pdf

[2] Duhem, P.: *The Aim and Structure of Physical Theory,* Princeton University Press, 9, July 1991; Popper, K.: *The Logic of Scientific Discovery,* Routledge, 29, March 2002.

[3] Balmer, K.: Wissenschaft. Der Freidenker, Jg. 29, Nr. 12, S. 93. Zitiert nach: Grebe-Ellis, J.: Grundzüge einer Phänomenologie der Polarisation. Berlin: Logos 2005, S. 33. (= Phänomenologie in der Naturwissenschaft 3). (In German. Quoted from Grebe-Ellis, J.)

[4] See the following contribution by Johannes Wirz: Goetheanistic Aspects in Dealing with COVID-19.

Johannes Wirz

Goetheanistic Aspects in Dealing with COVID-19

I started dealing with the issue of Corona when it was still bright and warm outside, case numbers and hospitalizations were low with hardly any deaths. At that time, I jotted down a sentence: 'We could possibly have had the best possible crisis science and crisis medicine for this Corona situation. But there is neither good science nor good medicine.'

Why did I come to this judgement? In the last ten months, thousands of publications have appeared on this topic and we know from the past—in terms of the history of science—that whenever something 'new' appears, an endless amount of research, thought and corresponding publications occur. Such a situation means that an overall concept, a theory or a law on the new topic has not yet been found—a challenge for every researcher. Yet I noticed that in matters concerning Corona, when it came to the measures and strategies, the same thing was done in every country, around the world! Detection, implementation of measures for physical and social distancing, masks and so on were carried out, and finally in the second phase, we heard that there is only one salvation—vaccines! How can it be that in a situation where a general picture does not yet even exist, that the same treatment and solution to the problem emerges everywhere? That was a mystery to me.

Today—at the end of November 2020—Switzerland has the highest number of cases it has had to date, with many hospitalizations and, tragically, more deaths. I have asked myself whether my original sentence still remains true for me—and I have to answer, yes. In the following, I will deal with evidence of the virus, look at the measures that have been taken and will also drop a few remarks on vaccination.

When scientists talk about the virus, they say: SARS-CoV-2 (which means: Severe Acute Respiratory Syndrome—Corona Virus—Type 2). When they talk about the disease, it is called COVID-19 (Covi for Coronavirus, the 'd' for disease / illness and 19 is the abbreviation for the year of its first discovery).

What is good science?

Before I get to the actual questions, let's first take a look at my own understanding of 'good science'. I will take the liberty of quoting Goethe here because he was not on the fringes of science in his time, but right in the middle of it. Two examples verify this: in the first issue of *Nature*, probably the most renowned scientific journal, Goethe's essay on nature was printed in full, in English of course. You can't imagine that today. That was in 1869, decades after his death. In 1882, Ernst Haeckel (1834-1919) wrote an obituary for Charles Darwin, who had died that year, and in it described Goethe as the great pioneer of the Darwinian theory of evolution—that is also unimaginable today. As Goetheanists, we are, after all, a tiny number on the fringe of a scientific enterprise that has itself become dubious.

What does Goethe say is needed for 'good' science? The first criterion is of course unquestionable: science must always be based on experience and observation. Simply spewing forth is not science. As a second criterion—overlooked by many Goetheanists—Goethe did not object to hypotheses, but expressed it in a metaphor:

> Hypotheses are scaffoldings erected around the building and are taken away when the building is completed; they are indispensable to the workman, only he must not mistake the scaffolding for the building. [J. W. von Goethe].[1]

A bigger problem occurs when projects are carried out that simply want to confirm a postulated hypothesis.

Goethe's descriptions of a scientific attitude, which can be found in his essay, *The Experiment as Mediator of Object and Subject* is interesting.[2] In it he says that the scientist should exercise caution, not put too much of his imagination into the project, not show any bias, actually perceive what is and what is not, not form judgements too hastily, but should also avoid envy and greed. This means postponing the formation of concepts and ideas. By contrast, collect the many physical phenomena and develop a comparative method in the organic or life sciences. In both cases, one must also try to look at what one is researching in the greatest possible, comprehensible overall context. This takes time and slows down theory building. In view of the more than 10,000 publications in the context of COVID-19, this would have meant that scientists discuss their theories, observations and descriptions, at length, with all kinds of experts in their fields, before publishing anything at all. So not asking themselves whether I am quicker than any of the others, but rather, what do my peers think about the issue/s? This view of good science is specific and unique, but it can, needs and has to be discussed today.

Finally, there are, as it were, soft criteria that have been completely lost sight of in science today: Goethe was convinced that when we look at life in nature—the plants, animals, human beings and the earth as a whole organism—we must always have two perspectives in mind. The outer perspective, for which observation is required. That is what today's science does. But it is also necessary to develop an inner perspective. The view from within does not mean measuring or counting, but rather involves relating myself as a living being to other living beings. The best description of this inner relationship would be to assign meaning to the sensory data identified from the outside. This inner aspect of living beings no longer exists in the conventional science of today.

Another point is central, which is also difficult for us because we so much like to feel in possession of the 'truth'. Goethe urged us to forego 'absolute truth'. Thus he wrote in a letter on 21 January 1832:

> So the human being ... cannot refrain from trying to drive the incomprehensible into a corner, until he is content and may willingly give into it.[3]

One last point, perhaps the most difficult, is completely frowned upon today: *ethics in cognition*:

> If the naturalist wishes to assert his right to free contemplation and observation, he makes it his duty to secure the rights to nature; only there, where it is free, will he be free, where it is bound by human statutes, will he also be fettered.[4]

This sounds a little puzzling at first; I will come back to it. First, I will turn to three topics in connection to COVID-19: the detection of the virus, the measures to contain the disease and vaccinations.

The Issue of the Detection of the Virus

The first thing that struck me is, on the one hand, the detailed investigation into the virus and the phylogeny of its origin. It is assumed with some plausibility that SARS-CoV-2 originated from bats. There is a problem with this. None of the corona viruses isolated from bats today would be able to infect human cells because they lack the so-called spike protein. Therefore, it is assumed that the virus must have come to humans via an intermediary host. Perhaps via the pangolin, which is sold in large numbers at wet markets in China. The second problem, however, is that thus far, this virus has not been found in the many pangolins that have been studied. So it is still unknown how CoV-2 infected humans. Experts suspect that the first spreading event occurred at a wet market in Wuhan, perhaps transmitted to a farmer or a worker.

Although all this is known, it is repeatedly claimed that it is a completely novel type of virus! This means that the evolutionary-biological and phylogenetic studies have not really been acknowledged. That is why problems arise when one wants to detect the virus. The PCR method (polymerase chain reaction), with which we are all familiar by name, is used to multiply the viral RNA. Without it, the minute amounts in the sample could not be detected. It is not known whether the method is that accurate that it only detects the CoV-2 virus or whether it does not sometimes also detect other corona viruses, of which there are about seven different types that have long infected humans and cause flu-like symptoms in winter. Many of us have been and get infected by them, and the infection is relatively mild. In the vast majority of cases, the immune system of those affected neutralizes them with antibodies or T cells. We will come back to this.

There is yet another problem using PCR detection. The virus genome has about 30,000 bases, nothing compared to that of the human genome, which has as many as three billion base pairs. In the PCR detection method, miniscule snippets with a length of about 120 bases are multiplied at two different locations in the genome of the virus. One gene locus provides the blueprint for a protein that can multiply the viral RNA (RNA polymerase), the other blueprint is for the so-called spike protein with which the virus attaches itself to the patient's target cells. The analysis does not confirm whether the viruses are intact or only viral debris, in the case of a positive result.

So we have a method that is precise, but yet still does not provide any exact information about the pathogen or the question of infection. The Swiss Federal Office of Public Health wrote that the detection of viral DNA is not a confirmation that the person, from whom the sample was taken, really harbours a virus.[5] So one wonders what the positive cases actually mean for the people tested every day. The positivity rate, i.e. the percentage of people tested positive (which is currently 20% in Switzerland) but is usually lower, says nothing about the number of people who have actually contracted the virus. Many who are infected by the virus don't even notice it. Others fall slightly ill and don't go to the GP. We only really know by those who have been hospitalized. Out of 20% who test positive, about 1,500 people per day in Switzerland during a peak phase, 3%, about 40 people, are admitted to hospital. Of these, only a tiny fraction tragically die from the disease. I describe this to point out that detecting the virus is not that easy.

I recently had a very heated conversation with a scientist friend who accused me of viewing everything incorrectly. He said that only by using the PCR test could the virus be detected quickly—and that this is important. Late at night, Swiss television broadcast an interview with a virologist who railed against the PCR sceptics by claiming that detection was 100% certain. Once again: whether the method detects the virus or only fragments of a dead virus (viral debris) remains open.

Dealing with the Pandemic

Now we come to the second topic and how to deal with the pandemic. One sure way of dealing with the pandemic would definitely be to find out how the number of sick people compares to the total number of people in a locality. The same with regard to the death rates. Personally, I would do without the current method of virus detection. Sure, one can disagree, but in my opinion, the less accurate rapid tests are probably cheaper because they detect at least one protein of the virus thus confirming its replication in the carrier. They are also easier to handle and less expensive than the PCR method.

As a result of the increasing number of positive cases from PCR testing, a whole range of measures are subsequently implemented, which we have all come to know: keep physical distance of at least 1.5 metres and as that isn't sufficient, wear a mask in public spaces. Then social distancing measures follow, i.e. no visiting or travelling, closure of restaurants, places of culture, shops and so on. Events like this one here tonight are still only with a maximum of 30 people, keeping distance and with masks. Despite these measures, one is amazed then to see how slowly the test case numbers are going down. There seems to be no understanding of the relationship between measures and outcome in terms of contagion, and the gaps in knowledge of the exact process of contagion. I suspect that the increase at the moment with the onset of winter is comparable to the increase in flu and colds in November, December and January each year. We will be able to observe whether and how quickly the case numbers decline, with the strictest measures, closures and lockdowns. I don't want to deny that something could and should be done. But I would also like to ask what the consequences of these measures imposed by the authorities are—not only here in the well-off, industrialized countries of Europe but also in the less well-off countries, in the so-called Third World or developing countries. From my perspective, the main concern is the risk groups, the elderly and people with pre-existing health conditions who should be protected, as suggested in the Great Barrington Declaration.

With us it is quite clear, and many have experienced it themselves, that these measures lead to strong feelings of loneliness and isolation in elderly people. If you are hard of hearing, for example, talking to your son or daughter through a window is simply not possible. I have heard from educationalists and psychologists that at a time when young children are learning to understand and love the world, they need closeness, touch and trust. By contrast, today things are the other way round. They find themselves in a world where distance, avoidance of contact and mistrust are seen as good. We now know that teaching with masks reduces the comprehensibility of what the teacher is saying. We have to learn that children not only listen to the spoken word with their ears, but also see speaking with their eyes. When this is missing, a problem arises. It has also been noted that children are no longer sure whether something is meant to be serious or funny, sad or cheerful, and so on, because the mask hides facial expression. So there are significant consequences associated with these measures.

Not to mention what we have learnt about people losing their jobs and very many small and micro-entrepreneurs losing their livelihoods, especially people working in the cultural sphere. If we look at the situation in its entire context in the sense of Goethe, we have to ask: What do we accept in order to protect ourselves against the virus, and how is what paid for and by whom? If we look beyond our own backyard to the larger world, things look even more dramatic to me. I just want to point out a few things: for example, in countries where day labourers feed their families on two to four dollars a day, huge levels of poverty have occurred. It is estimated that as a result of Corona, 130 million more children today are starving than before the crisis.[6] You have to try and imagine that! We also need to understand that since Corona, the treatment of tuberculosis, malaria and HIV have decreased massively. Experts estimate—you can read about this in *Nature, International Journal of Science*[7], where Goethe's article was also published—that about two and a half million people have died from these infectious diseases as a result. The experts point out that the entire further development and production of cures or vaccines has of course been completely scaled down in the current situation and will probably only be partially resumed in the future.

So here too, in the sense of Goethe: if we look at the whole context, can we really enable one thing to happen and allow the other not to happen? I don't have an answer to that. However, if I look at the statistics for tuberculosis, malaria and AIDS, they are probably far more dramatic than COVID-19, globally.

The Question of Vaccinating Against COVID-19

I would like to start the topic of vaccination with some good news. In a study of more than 1,000 people, antibodies against the CoV-2 virus were looked for in people who had not been in contact with it[8]; the results were negative. It is important to know that humans have two different acquired immune systems. One is called humoral, which creates antibodies that float in the blood and lymph. About 6 to 12 months after an infection, they are no longer detectable, in most cases.

In addition, there is the cellular immune response, in which so-called T cells are able to recognize and subsequently destroy the cells affected by the virus. In the laboratory experiment, it was found in the above-mentioned study that 35% of these people without prior COVID-19 disease had T cells that recognized and neutralized the current coronavirus. It is assumed that these people had previously come into contact with another coronavirus, had overcome it and the immune system had produced T cells that now also recognize SARS-CoV-2.

So SARS-CoV-2 cannot be as novel as specialists claim. If the results prove to be true, the population would already have achieved a major step towards herd immunity.

Now to the vaccines. There are a number of ways to produce vaccines. One classic method is the way vaccines against influenza viruses are produced. The viruses are grown in chicken eggs, heat-inactivated and injected. We can also work with antibodies that recognize structures on the viral envelope and can destroy them. The focus now has however been on a single method: the so-called RNA vaccines. Why? It is the only way to produce millions or billions of vaccine doses in a short time. It is nonetheless important to know that although this method has been known since 1994, it has never been approved for the production of a vaccine for humans. This is simply because years of testing are needed to determine how long such vaccine protection lasts, and what the medium-term to long-term side effects to be expected are. Despite this, a number of these novel vaccines are now being used in humans, without adequate testing.

It struck me that there has been no mention of compulsory vaccination to date. Are the authorities not convinced of their effect, or are they serious about freedom of choice? No, there is another, legal reason. If compulsory vaccination were enshrined into law, the authorities would have to assume liability for any damage caused by them. Without compulsory vaccination, the companies that produce the vaccines would be liable.[9] Not unexpectedly, the vaccine companies wanted to shift this liability away from themselves. It seems similar to the nuclear power plants' liability issues. They would never have been built if the state had not been prepared to cap the cost of liability at two billion francs (in Switzerland); in the event of a worst-case scenario, a worst-case accident, they knew full well that this cost could be more than ten times higher. With this new vaccine, I have the impression—this already sounds like conspiracy theory, of course, I confess—that the COVID-19 vaccines, which are portrayed as being the only alternative, are heading in a similar direction as the nuclear power plants in terms of liability for possible consequential costs in the event of severe and fatal side effects.

However, there is something else that has caused me great concern. The RNA-based vaccines are supposed to trigger an immune reaction in the vaccinated. This reaction is very strong in children and young people, in the majority of whom COVID-19 disease is harmless. In old people, in whom the course of the disease can be severe, the immune response is weak or fails completely. So it would benefit people who don't need it, but not all those who urgently need it. So I dare to doubt the point of using it, especially for the elderly. It is not yet clear how long the vaccine-induced protection will last and whether it will still be effective against SARS-CoV-2 viruses, which will mutate over time like the flu viruses. Limiting the therapeutic view of the virus and immune system, which always reminds one a little of warfare—there's the enemy, here's the defence—would have to be complemented here by a 'peripheral view' in the sense of Goethe's broader perspective. For example, what behavioural patterns and factors lead to mild, unproblematic courses of disease in infected patients? Which therapies could be used to strengthen powers of self-healing?

What next?

So much for my sceptical comments on the detection of the virus, the measures taken and the vaccine. If we now take another look at the larger context of the phenomena of this corona infectious disease, we cannot avoid considering not only economic, social and societal problems, but also ecological ones. SARS-CoV-2 is a zoonosis. This means that the virus is transmitted from an animal to a human being. Just as Ebola in Africa is also a zoonosis, in both cases, in Africa and in China, the transmission to humans took place after massive destruction of the natural habitats of bats. These animals have no problem with these viruses. Maybe they get sick and recover. If all the trees where bats live are cut down, they naturally gravitate to environments where there are still trees: in settlements, villages and cities.

When you look at the whole context, both socially in human beings and ecologically in relation to the environment, you come up with a stirring image: the health of the human being is closely connected to the health of the earth and, conversely: the bad health (disease) of the earth is closely connected to bad health (disease) of human beings. Perhaps such images only speak to anthroposophists.

As a result, of course, the question immediately arises, what actually has to be done now, in this moment of need? How can actions of those responsible be guided in a different direction from those we are accustomed to? I learned from Rudolf Steiner that we need a cultural transformation. The primary principle of maximizing the happiness of individual needs to be transformed into a culture where the primary principle is to minimize the suffering of others[10]. This is a major development in consciousness that we would have to take on. It seems to me that we have to pay attention to what the consequences of Corona are, not only in small, rich Europe, but in the wider world. In terms of their ecological consequences, if we consider why these pandemics have occurred at all, and will occur in the future, it seems to me that such a transformation is inescapable. I think movements and efforts would have to start in this direction, which could even begin now, in these times of crisis.

A second point concerns the renunciation of absolute truth, which Goethe poetically formulated, as mentioned above:

> So the human being ... cannot refrain from trying to drive the incomprehensible into a corner, until he is content and may willingly give into it.[11]

What does it take to deal with the incomprehensible? Which scientist today would dare to say: Stop! The drive into the 'narrow' is still going on: in quantum physics with the construction of ever larger detectors to discover ever smaller and even shorter-lived 'particles' for the unravelling of the mystery of matter. Or in biology, with the perfection of methods to decode ever larger quantities of genomic sequences with electron microscopes, with which individual proteins can be made visible close to absolute thermal zero.

Giving into the inexplicable has, I believe, a lot to do with dismantling hypotheses, the scaffolding, and reflecting on what exactly has been gained and lost with the methods of modern natural science. This is perhaps the moment when one can, and perhaps must, renounce further research into the manifest world and begin what Rudolf Steiner calls the path into spiritual science.

For this, we would have to go beyond normal consciousness or sense-based knowledge and develop thinking into an organ of perception.

This brings me to my last point, a new ethics of science. I recall Goethe's saying at the beginning of my contribution:

> If the naturalist wishes to assert his right to free contemplation and observation, he makes it his duty to secure the rights to nature; only there, where it is free, will he be free, where it is bound by human statutes, will he also be fettered.[12]

If I use a method in which I determine in advance what nature has to tell me, a method demanded by Francis Bacon (1561-1626), I subject nature to torture; the screws are tightened, so to speak. In this way, a yes or no can be wrested from it, but then, not only is the freedom of expression robbed from the things under investigation, the living beings, plants and animals, but the scientists themselves are robbed of their freedom. We are not only closely connected to the world as a living being among living beings, but also through our consciousness and our way of thinking about the world.

No one has captured Goethe's scientific ethos better than Adolf Portmann (1897-1982), the Basel zoologist who, in my eyes, is one of the greatest non-anthroposophical Goetheanists:

> Goethe lived an attitude which was guided by *reverence* as his central motive and he carried out renunciation, the renunciation of destructive intervention. In the study of nature, reverence does not just preach a kind of non-violence, it fulfils, (...) from which no one can withhold the greatest innermost respect.[13]

When one reads and reflects on these words, the importance of non-violence in dealing with nature becomes clear. In this way, we learn to look at the world and at our fellow human beings with new eyes. This could be the beginning of entering into a different relationship with nature, and hopefully with human beings as well; relationships based on respect, love and possibly also peace. As important as it is to remain capable of action in the current situation, it is equally important to learn and develop this attitude as a condition for good natural science.

Revised version of a lecture given at the Goetheanum on 23 November 2020.

[1] Goethe, J.W. (1897): In :R. Steiner, *Goethe's Conception of the World*, Personality and View of the World. Translation, Collison (GA 006).

[2] Goethe, J.W., *The Experiment as Mediator of Object and Subject*, essay 1792. Translated by Craig Holdrege. The Nature Institute, 2010.

[3] Goethe, J. W. (1832). In: Trunz, E., Ed.: Goethes Werke. HAB IV, S. 467; Beck.. [Goethe's Works, HAB IV] Quote translated by C. Howard.

[4] Goethe, J.W. (1834): Goethes Werke. Einundfünfzigster Band. [Goethe's Works. Volume 51] Quote translated by C. Howard.

[5] The FOPH (Federal Office of Public Health, Switzerland) had this statement on its website in the summer of 2020. After this lecture, a member of the audience drew my attention to the fact that the FOPH claimed in November 2020, without giving any reasons, that the PCR test provides reliable proof of the virus.

[6] Nordling, L. (2020); 'Pandemic of hunger.' *Nature* 12. October 2020. Accessed: 23 June 2021.
https://www.nature.com/immersive/d41586-020-02848-7/index.html

[7] Anonymous (2020): '*AIDS, malaria and tuberculosis are surging.* Efforts to defeat the coronavirus have fuelled a rise in other infectious diseases. Urgent action is needed to avert a catastrophe.' *Nature, International Journal of Science,* 13 August 2020. Accessed: 23 June 2021.

https://media.nature.com/original/magazine-assets/d41586-020-02334-0/d41586-020-02334-0.pdf

8   Nelde, A. et al. (2020): *SARS-CoV-2 T-cell epitopes define heterologous and COVID-19-induced T-cell recognition*. Research Square Preprint; https://doi.org/10.21203/rs.3.rs-35331/v1

9   Maier, M. (2020): Corona-Impfung: Wer haftet für mögliche Schäden? Berlinerzeitung 26. August. [German article in the newspaper, Berliner-Zeitung. Corona Vaccination: Who is liable for possible damages?]. Accessed from: https://www.berliner-zeitung.de/wirtschaft-verantwortung/corona-impfung-wer-zahlt-fuer-moegliche-schaeden-li.101215

10 Steiner, R. (1918): *The Work of the Angels In Man's Astral Body*. (GA 182)

11 See endnote 3.

12 See endnote 4.

13 Portmann, Adolf (1956): 'Goethes Naturforschung', *Biologie und Geist*, p.273, Taschenbuch 124. ['Goethe's Nature Research'. In: Biology and Spirit]. Quote translated by C. Howard.

Stefan Hasler / Ueli Hurter

Individual Responsibility in Corona Times

Ueli Hurter: We Swiss actually take great pleasure in any form of individual responsibility, because we think we understand it well and can live it. As the Corona time has progressed, we have of course become aware that individual responsibility, in relation to the pandemic, is also a sensitive issue in socio-political terms; individual responsibility versus collective or state responsibility. We believe this is an important topic for discussion, perhaps even for debate. We will not go into this here, but we would like all the more to encourage people to sense the dimensionality, the rationale and the possible meaning of what individual responsibility can be and to make it clear for ourselves. We would like to strengthen individual responsibility in social discourse as opposition to collective or state responsibility.

A brief clarification of terms is important at the beginning. We generally use individual responsibility and self-responsibility synonymously and make no distinction. Individual responsibility is not meant here as a kind of egoism, for myself, but means in our sense: firstly, I am responsible for myself. Secondly, I am also responsible for my immediate social environment, due to my sense of individual responsibility. Thirdly, I am also jointly responsible for the good of the common whole. A second clarification of terms: what does 'self' (German, *Eigen*) mean? First of all, of course, it means a person, an individual—or now, in the context of the Corona crisis, a citizen. However, we mean—and this has become experience in recent months—that there is also individual responsibility on the part of an institution, for example, the Goetheanum. There are also regional authorities for larger regions, such as the cantons in Switzerland, which are autonomous. This is perceived quite differently in the canton Solothurn than our immediate neighbour, Basel-Land, and slightly further away, Basel-Stadt. Individual responsibility therefore refers not only to a person, but to various social entities as well, entities on their own, so to speak, which need to act in a situation like the Corona pandemic and take responsibility themselves. Stefan Hasler will describe in the form of a report, or as a kind of practical research, how we have been able to manifest this individual responsibility in and for the Goetheanum in recent months.

Practice Report of the Goetheanum

Stefan Hasler: I am speaking here as part of a three-member management team at the Goetheanum with around 160 employees who have been very practically involved with this issue every day over the last months. At the beginning, this lockdown took us all completely by surprise. That it is at all conceivable to close down a public place like the Goetheanum, which is open seven days a week from 8am to 10pm with events, meetings and conferences, simply through official decree, has not happened here since the Second World War and no one, before now, would have thought it possible.

The situation in Switzerland then progressed politically, step by step, after the Federal Council took on the overall responsibility. This responsibility was handed over to the regional authorities, that is, to the cantons and then on to the local municipalities. The clear stipulation was that the individual institutions were responsible for compliance with hygiene regulations in each case. Each institution was required to have an individual protection/safety concept, but this did not require approval by the local authorities. No local authority wanted to see a protection/safety concept, because the responsibility could not simply be handed to the local authority, so the institution became responsible within a given framework to find something that worked.

This gives the management here a different feeling when the responsibility lies with the institution itself. This even includes possible flexibility for changing the concept every month, for varying it, for inventing something new and for adapting it adequately to the immediate situation. We became acutely aware that with an institution such as the Goetheanum, we are a public place and therefore part of society, of this community, of this canton and of this country.

Culturally speaking, we were very happy that we were able to perform, despite a postponement, Goethe's *Faust*, parts 1 & 2 in the summer of 2020, with 9-hour performances, on each occasion with around 550 people in the hall. We were equally happy to be able to do an additional performance in the autumn, with 650 people in the hall, just on the last weekend before having to return to a restriction of 30 people per event. As far as we are aware, no COVID-19 infections occurred at any of these performances due to our safety concept.

For Rudolf Steiner's 3 Mystery Dramas at Christmas, which were finally subject to a complete ban on events, a concept was developed that the dramas would be performed in such a way that 30 people could attend each performance, meaning a total of 90 spectators would be in the building at the same time, at three parallel venues, so that each scene would be performed three times and the audience would then simply move on to the next venue after an hour, and then again to the next venue after an hour. In the end, all that was left to do was, one, an online production of rehearsal workshops with the artists, which lasted several hours and will hopefully serve to attract large audiences to future live performances, and two, an internal complete performance without an audience at all. A very clear statement from all the cultural workers at the Goetheanum, and from the Goetheanum Leadership as well, was made: we want to perform as much as possible, everywhere, creating anew so that as much culture as possible actually takes place.

Two experiences are crucial for us in this regard. As a public place, we have wanted to re-examine what is possible within the current situation and be in active exchange with all those in charge. This was actively done in 2020, from the village police officer to the mayor, from the Office of Culture in Solothurn to the responsible cantonal councillor and president of the cantonal council. Always with the same aim: how do we deal with the situation in the best possible way? It has become clear how the Goetheanum, like other institutions, has long since become an important cultural institution in the context of the wider regional environment.

Even the Goetheanum has not been able to completely avoid individual COVID-19 cases occurring here as a matter of course, for instance, during the large medical conference held in September 2020. On the middle day, when 800 people were gathered here, a participant developed symptoms, left and two days later, while the conference was still in progress, learned of her positive test result, which she immediately communicated to us. On Friday evening at seven, after police mediation, we reached the responsible cantonal doctor ten minutes later and were able to clarify in conversation how we could continue the event in the evening at eight, by making a clear announcement. The matter was filed in writing the next day and the following week, and fortunately, we were able to proceed without any further spread. Likewise in late autumn, the students' residence of the Goetheanum had to be quarantined for 14 days because several students tested positive and there were some minor cases of illness. The decisive factor in each case was the discussion with those specifically responsible in the community and in the canton, and the responsibility was thus positively shared in the concrete situation through person to person contact.

Ueli Hurter: I would like to emphasize once again for this report what the basic principles in Switzerland actually are. Sometimes, a regulation is passed that is quite invasive for us, that really gets under one's skin. One is then in the position of having to implement it, out of one's own sense of individual responsibility, and this will not be controlled from above. We did have one visit during which two inspectors looked around the site, never checked our safety protocol, but made some suggestions for adjustments, which we did.

I had a comparable experience on my farm because we had planned an event with 200 people when suddenly a colleague pointed out to me that such an event had to be registered at least 14 days ahead of time. It was now five days before including a weekend. I now assumed that the event could not take place, but on the Monday morning, I quickly called the canton of Neuchâtel, and apologized, saying that all the invitations had already been sent out etc. The lady responded very calmly: 'I'm sorry,' she said. 'Now tell me, what do you have in mind?' I told her and then she said, 'No problem. Make your own safety concept. You are responsible for carrying it out and everything will be fine.' You feel a sense of responsibility that surges up your spine. You feel that you can stand upright again, because you have not been enslaved or pressed into any kind of mould, but have become a co-contributor to what happens in public places, out of individual responsibility.

The State as a Self-Help Organization

Stefan Hasler: It is very important for us, especially as a public place of culture, to ensure that culture gets the space it needs in the larger discussion. Basel's cultural workers have written a public letter to the Grand Council in Basel and to the Federal Council in Bern, in which they criticize that shopping and tourism are allowed for thousands of people and cultural events with clear safety concepts are not possible for smaller groups of people. We co-signed this letter. Many anthroposophical cultural workers were present at the demonstration. It is a matter of great importance to us, especially in a democracy, to bring this aspect clearly to the foreground.

In this regard, a formulation, which is characteristic for Switzerland, was made by the jurist Professor Christian Brückner, who put it in a nutshell: 'In our cooperative understanding of the state, the Swiss understand the state as a self-help organization.' This is to be taken literally, because the

Swiss 'confederation oath' (German: *Eid-Genossenschaft*, English: confederation) results from the fact that everyone is aware that what the individual cannot achieve alone, and is only possible through a group, is actually the union of cooperation that wants to achieve this or that goal. The confederation is therefore a self-help organization for everything that the individual cannot achieve alone. Switzerland as a state, as it is known in most other countries, has never existed in this sense and does not exist except in the sense of a self-help organization. This is why there is such a pronounced culture of autonomous cantons, municipalities, individual institutions and of individual citizens.

We think that what we have experienced so far during Corona, more or less corresponds to the Swiss way, in the sense that the Federal Council set a relatively soft regime, and everyone is individually responsible to a high degree. One has the feeling that it has been possible up to now, and we are talking about the period to the end of November 2020, that we have been able to manage this individual responsibility in a targeted manner and have also achieved something. Of course, the situation is still open. And there are already voices complaining. 'The virus is totalitarian and has produced a superstate' is what a Swiss wrote in a Swiss newspaper. So even what we have achieved so far, for the Swiss' sense of right, has already led to a superstate, so to speak, and overcoming the virus also means overcoming the state of emergency of the superstate.

How is the zeitgeist speaking through these events?

Stefan Hasler: We would now like to take a look at another quality of individual responsibility. We live in a situation that deprives us of the mobility that has become so natural to us with local and long-distance travel, and which ties us physically to one or other place. Nowadays, I am happy to go to Basel from time to time—comparable to the '15 km radius rule' in Germany. Worldwide, we humans are being forced to rest, so to speak! We have always had big travel plans for the future, knew long ahead where we would go for our next summer holiday, and were still inwardly consumed by what we had experienced just three months before.

With all the mobility we were used to, the present feels a little like something we have lost. At the moment, hardly anyone is thinking about what they will do during their next summer holiday. The question does not seem to arise at all. Today, I'm not even sure how I understood the world 'back then', before the pandemic broke out. From a spiritual point of view, we are challenged to live much more strongly in the 'now' and to face it. Whether we actually do so is a second question. But the spirit of the times, the zeitgeist, calls on us to live in the present, in a new and flexible way.

On the spiritual level, I observe two things: on the one hand, I often see an unvarnished honesty in social interaction that did not exist before—tensions that were rather latent in collegial experience in recent years, have now surfaced. A certain veil that had been spread over our lives is being ruthlessly unveiled by the Corona pandemic. Questions like: Where do I stand? What do I want? Do I accept the challenges and rise up to them? This honesty arising out of our spiritual nature strikes me. And on the other hand, I experience a question regarding the quality of the meetings I have with my worldwide contacts. A new, different kind of transparency and intensity is emerging; an openness to certain experiences and perceptions that was not possible before.

Being calm in the body; being present in the soul; and being transparent and honest in the spirit: if we look at it in this way, at what the spirit of the time is demanding of us, I see it as something comparable to the meditation posture. Prerequisites for meditation are to bring myself to calm and rest in my body, to be completely present in the moment in my soul, and to be as transparent and honest as possible with myself in my spirit.

Ueli Hurter: I would also like to take this three-step approach that Stefan Hasler has evoked for the individual human being in terms of our bodily, soul and spiritual dimensions, but now more in relation to the social domain. When the body comes to rest in a certain place, then this place is mine and I am responsible for this place. The vast earth on which I expansively moved previously, now becomes a mosaic of many individual human places in the sense just described. In this contemplation, my body is connected to a piece of earth; I nourish myself from it and in the quality of my nourishment, I take responsibility for this piece of land.

I have to breathe. My individual responsibility is more in the sense that I share air with everyone present, because we breathe the same air. This touches here upon the atmosphere and the climate crisis. Here it is about my surroundings, because I have to live somewhere, for example. This raises questions about the materials and the energy I use and for which I am responsible. Here, too, the question is: What is the relationship between what is mine to the whole?

Thus, one could say that the earth as a whole is represented, is present in every little bit of it. To the extent that I take responsibility for my place, I can help solve global problems. The motto 'Think globally, act locally' is always ingenious. Don't think locally and act globally, but the other way round! And the pandemic in particular shows us that. As it is a pandemic, it affects all people. We are actually all subjects of this pandemic, or objects, and at the same time, everyone is standing at a place where this pandemic can be directly encountered. Do we need a global or a state strategy to combat it? I would say, if so, then it needs to be based on a strategy of individual responsibility. So much for the reference to physical space.

Now to the spiritual aspect. As Stefan Hasler explained, I am currently living in the 'now' and am challenged in the 'now', the present. If I extend this to something larger, then for me it means that we become aware of our responsibility as a generation in the wider context. Because seen in a larger rhythm over time, the present can be equated with a generation. What is the mission of this, our generation of today? We have different generations living side by side in social life, but each generation must now ask of itself: What is my contribution and how can I enter into a generational contract? We have become very aware of this through the youth movement called Fridays for Future, which campaigns for our climate. This is a generational contract! Do we honour it? Are we able and willing to do so? People are already talking about an agriculture that is fit for our grandchildren or a climate policy that is suitable for an epoch.

There are many things I cannot do alone, but a single generation can. Nevertheless, what I can do alone, of course I should do. This is about modesty in our aspiration, but with a certain determination and resolve in implementation. The individual responsibility of each generation, but also of the individual, must be established afresh. After the totalitarianism of the twentieth century, after the end of the Cold War, after the consumerist frenzy of globalization, and in this sense, Corona is a wake-up call—from the point of view of the spirit of our times—to assume individual responsibility, which we know must be manifested differently in economic life, in political life and in spiritual life.

With regard to the third aspect, the spiritual dimension, honesty and transparency were mentioned here. When I enter into conversation with myself under these conditions, when I struggle with myself, I discover an 'I' within myself. It is neither material, therefore not an object for natural science, nor a juridical subject in a state or in social science, but is a fundamental inner experience within me, based solely on itself. And this 'I' that I experience in myself—this is perhaps the most primary experience of what spirit is—that my 'I' has an individual spiritual imprint. *I am spirit.* And this 'I' seeks 'you', finds you, and this you is 'I' experienced from within you in the same way that I am 'I', as experienced within myself. *You too are spirit.* This is another human being, regardless of whether it is a ten-year-old pupil or a ninety-year-old man. The same applies to everyone: individual responsibility is founded in my being, founded my being, and opens up the world to me as a being, in a world of beings, a being-world.

We—dear reader—are well aware that this is the limited view of two Swiss people. At this time and place, we wanted to speak out of the very concrete experiences that we have had here at the Goetheanum.

A slightly edited and shortened written version of the lecture given at the Goetheanum on 30 November 2020, which included a live sketch by Daniel Kehlmann on the corona situation and a characterization of the different ways in which the corona pandemic is being dealt with in different European countries.

Florian Osswald / Claus-Peter Röh

Digital Challenges in Education

We have decided to engage in a dialogue to examine the digital challenges in education from a specific perspective. This problem is currently a pressing issue in schools all over the world due to the Corona situation. We will begin here with the archetype of Waldorf education methodology of 1919, which back then meant a general radical change in educational approach. Are we now once again in the midst of an epochal change? What are the digital challenges of today actually about? How does mechanization and digitalization relate to Waldorf education methodology? And the core underlying question however is, how can we deal with digital media in a more humanly oriented way?

I would like to ask Claus-Peter Röh a first question:

What was the nature of this paradigm shift of the Waldorf School in 1919, being the first school that tried to bring this shift into the world? What would you say is its quintessential quality? Paradigm shift sounds good—or was it only just about an improvement in methodology or a small change in the overall question of approach in education?

Claus-Peter Röh: The impulse of Waldorf education is something like an eversion or inversion of the entire educational approach. The outcome could no longer be compared to the existing school system. Such a paradigm shift does not usually originate from the outside, but usually through people who initiate it and who are then able to realize it together with and for others. Isn't it basically so today that new questions are knocking at the door of schools, including Waldorf schools, now just three generations after the founding of Waldorf education? So if a paradigm shift can only be initiated by people, then we are being asked to develop something new out of this situation. Of course, this is not yet there or even ready.

I would like to use a particular incident to approach the quality of the paradigm shift in education in 1919—through an artistic contemporary of Rudolf Steiner. Perhaps a certain inner field of symmetry can immediately be sensed between the two people who—without knowing about each other—were actually asking and researching the same thing. In April 1914, an artist travelled to Tunis. It is the famous trip to Tunis which Paul Klee (1879-1940) made together with August Macke (1887-1914) and another artist. In terms of art history, this Tunis journey overturned everything that had previously been developed in expressionism and resulted in a new look at the world. Klee then wrote sentences like: 'Colour has now got me!' and 'Colour and I are one!'. He dissolved the traditional forms of his art, the figures, the outer contours, and then found his way to cubism through colour. Simultaneously—this now at the same time as the founding of the Waldorf School—he was very intensely occupied with the image of the human being.

He studied movement in particular, in himself, in others, and noticed: when you explore the movement of a human being and his physical form, you would basically always have to have a dancer around you in order to not only feel something yourself, but to recognize the movement in the space between you as well. He discovered something about the form of the human being and engaged with the head of the human being—in the form of a gesture—and did something like this:

Beyond that, he now asked a next question: How is this head connected to the whole human being, to movement, to the body, to the whole earthly being? Then he drew a second gesture:

1

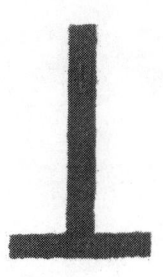

The head is not alone in the world

Klee drew, but also spoke to the head: 'You head, you are so terribly important at this time, a kind of Nuremberg Funnel, education from head to head, passing on knowledge from one head to another head, as quickly as possible, ….' In the exhibition *Bewegte Bilder* (Pictures in Motion) by Paul Klee in Bern in 2016/17, I found this particular sentence that, as a gesture, pretty much contains in it everything new from the time around 1919. Klee responded to the head, saying:

> *The head is not alone in the world!*

We could just debate this sentence alone. Now, along comes an artist—for the educator is always an artist and we will therefore still have to expand the concept of art in education—who describes how different these two elements, the 'head' and the 'rest of the body' with its movements, are. The head is not alone, but is connected to the rest of the human body, which abides by entirely different laws. Klee describes this as a dramatic conflict, and the older a person gets, the freer he becomes in his thinking, the more dramatic this conflict becomes.

> From the opposing bondage of its pure, spiritual capacity to the earthly bondage of the whole body, a more or less tragic conflict arises, which takes on more intense forms as the freedom of thought increases. What can the body do in relation to the spirit? [Paul Klee][1]

The drawing here hints at a motif of the later cubist period of around 1937. Klee, of course, does not leave it at that.

He asks about the connection between these two opposing poles:

What can the body do in relation to the spirit?

—and then draws as follows:

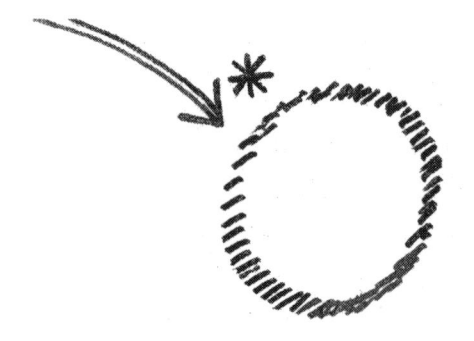

Klee describes this connection as a tragic conflict.

What can this body do in relation to the spirit?

Rudolf Steiner (1861-1925) was asked by the entrepreneur Emil Molt (1876-1936) to found a school based on the new insights arising out of anthroposophy, which seeks to unite the spiritual, the soul and the body. Rudolf Steiner arrived at a basic principle that he characterized no less dramatically than Paul Klee had by describing a form, a structure of the human being. This is the head, which has the tendency of being 'gifted' in life out of its pre-natal cosmic origins.

When a child is born, one is surprised at how fully formed the child's head is in relation to everything else. This is what the child brings with it. From all that happened before birth, one could draw a kind of arrow coming from the past to the moment of birth. This is then the one pole, the tendency to roundness, the spherical and the cosmic.

But the other pole also claims its rights, but at the very beginning, it is still very young.

Between these two poles Rudolf Steiner characterizes the human being: this cosmic form of the head, this earthly essence of the limbs. When a child stands up, as a rule sometime around the first year of life, the legs and feet transform, because they now have to place themselves within the gravitational forces of the earth. This young human being is now carried by its own legs. The tender feet of the infant transform in the first year of life so that they become feet fit for the earth, become earth feet.

Rudolf Steiner characterizes this polarity at the same time as Paul Klee: 'The human being is actually enclosed, wedged in'[2] between the spherical, cosmic tendency of the head and the earthly alignment of the limbs. This polarity seeks a solution, calls for an interconnector, for a third element.

Because there is a middle sphere connecting the two poles, this will be decisive in our further considerations. This is where the transformation of the old educational parameters takes place! Now the old aims no longer apply, this bringing of knowledge from one head to another head as quickly as possible. From now on, a new methodology begins that involves this middle realm. Although, there remain certain publishers, still today, whose books are about the great wealth of knowledge and how to get it into our heads. This approach is simply still strongly present. Rudolf Steiner, on the other hand, in his educational approach, set a path as a process of learning: that which I take in as knowledge has to first pass through the whole human being, through the senses, through the experiencing of it, through the doing of it oneself.

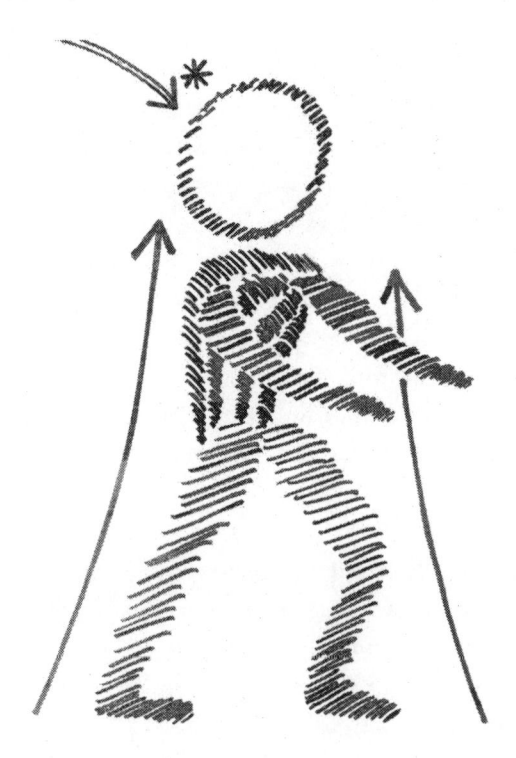

Such is the path in the new educational paradigm and as a result a different, more enlivened thinking ensues quite naturally. Can you notice this in people when they speak, what path their thinking has taken before? Or how it might have been formed in a completely abstract way?

What is a kilometre?

In November 2020, Remo H. Largo (born 1943) passed away, a great Swiss paediatrician and author who gave guidance to many parents through his books and lectures. He always emphasized that every young person is an individual. So be careful with all standardized ideas, he warned, that are simply introduced into the upbringing of young people. It does this individual child no good. In an article in 2018, in which he once again took a stand against certain curriculum developments in Switzerland, as a kind of testament, he said, 'If you standardize children, impose something on them or expect something from them that does not at all correspond to their inner being, they will then either withdraw or over-adapt. A child is helpless in the face of such standardized demands that it somehow just tries to fulfil, but not out of its own inner impulses. Children become over-adapted beings!'[3] This was an article of an interview with him in the *Basler Zeitung*. Then the journalist asked a question. In his opinion, what would a school be like that really points the way to the future? What would make up such a school? Remo Largo answered: 'Learning is only possible when it is concrete, when it is based on concrete experiences—and only when it is stimulated out of self-determination. That alone is learning.' The journalist asked: 'Do you have an example of this?' His answer: 'My third-grade teacher said in a maths class, "Tomorrow, all of you bring a metre-measuring stick with you." And the whole third-grade class took great pleasure in placing these metre-long sticks together until they had measured out a 100 metres, and then repeated the same thing ten times over. At the end, they experienced with the joy of discovery, the distance covered and what a kilometre is: 1,000 times the metre ruler.' Remo Largo added: 'To this day, I know exactly what a kilometre is! That is learning!' He became radical here and emphasized that all this writing of numbers on sheets of paper and repositioning them with symbols and so on, all this will never provide a real sense of space!

In conclusion, we can thus say: every concept has a personal history, for each one of us. Today, this is also called 'framing'. This implies that the terms I use, whether in life, in society or in the COVID crisis, have a 'framing' and it is good for me to know how *my terms* came into being in my story or how yours came into being. From this also follows one of the goals in education: to enable lively conceptualization, through experiential processes.

Now I'd like to ask Florian Osswald how this particular methodology relates to the development of technology or to the handling of the digital? How do we experience this through technology?

Technology Belongs to Our Lives

Florian Osswald: Claus-Peter Röh has described how human beings live in their threefold bodily structure with its head, limbs and middle area, and that in addition today, we all have manifold experiences in the use of digital media. The fact is that when I use such a medium, my body remains quite still, while my head is somehow very active. So when I sit at my computer, my head is active and my lower body slowly falls asleep and my feet get cold. These are two areas that are separated from each other. That's the first primary phenomenon to grasp when using media technology. There is obviously a separation between head and body and the question is: What is then happening in our middle realm? What is there to balance these polarities? If we as teachers are specifically trying to work with this middle sphere at school, it doesn't help if we use media a lot—or, we have to become more actively involved in how to deal with it.

In our everyday lives, we all need media nowadays: we use our smartphones every day, maybe also voice messaging, translation programs, and so on. We love streaming films or listening to music, and we have many ways of communicating using different kinds of platforms on social media. There is a huge abundance of it, whereas we older people might still just use emails; something that young people don't seem to use much any longer.

Then questions arise for the pupils at school: What happens to everything I put into this machine? Questions like: What about data protection, or increasing misuse? Concerns about security arise quickly, and it has become very tangible how the economic sphere has a very strong digital influence and presence. Artificial Intelligence is being used extensively. What about robotics that have been developed and are now in use? Online trading which has grown enormously during Corona times? In a short time, an incredible number of new things have emerged, that for a long time now, have been a matter of course in our everyday lives.

And the young generation is now growing up directly into this world. This means that it is not the same thing as it is for those of us who knew the world differently. The world has changed. For these young people today, it is their total reality. As parents and teachers, we have to put ourselves into this situation, into their shoes, if we want to understand these young people at all, because this is their reality. They don't know it any other way. There are smartphones, you use them, so you can't just say: Well, they are not good.

A colleague just told us of a lovely example from Kenya, where everyone—even more than here—has a smartphone. This is a blessing, she said, because all payment transactions are handled via smartphones. Now, for example, you can receive your salary directly on your smartphone, and it's wonderful, because corruption has been curbed that way. Before, the wages came and the two or three people in between always kept some money for themselves. Now it's all neatly regulated. So the feeling that the smartphone has changed life and made it safer, lives strongly there. 'I get my wage', even the women do, whereas before, the men had already taken some of it before she received any.

How do the pupils arrive at knowledge?

When we look at what this world is today, we have an interesting split or tension in the question: How do I know the world and how do I act in it? That is one of the big questions we have in education: What do we know and, then, how do we act?

I once asked my mother if she knew how her sewing machine worked. As a little lad, I had a detailed look beforehand and then I asked: 'Can you tell me how it works with the thread that goes up and down there?' My mother said she had no idea how it worked. I was very proud of myself because I knew. She then asked me if I could sew, and I had to confess that I could not. She could sew very well, but didn't know how the thread made a stitch. She was never interested.

How can we arrive at a realization with our pupils that this is a world that gives me inner security, when in relation to a process, I say: This is how it is. Understanding is a basic right! It is important that we do this together with the children; that we open up the world and show them what is out there; how diverse this world is, in order to then promote a development of knowledge, to a recognition of how everything is interconnected. This then leads back to a new, experience-saturated understanding. We try, out of diverse experiences that are as rich and broad as possible, to help the formation of concepts.

But the greatest concept for their development is that of the human being itself. What are we actually? That is the big question we put to the growing human being, but it is a question that everyone can always ask themselves anew: What is the human being? How can the image of the human being be formed over the many years of being together with the children? How can this be taken up again and again in keeping with the age of the child, because the educational questions of a 15-year-old person are completely different to those of a 7-8 year old child?

However, there is also a great potential for tension here, especially in relation to the world of machines. These are, after all, put to us as questions from life itself today, and you even have to study the subject of 'action' robots at school. So today, of course, Artificial Intelligence also forms part of current knowledge. You cannot avoid dealing with it.

Can I get by without my smartphone for two weeks?

The big question is, of course, when does one start with this? Can this young person handle these sophisticated devices at all and at what point? This is one of the most important questions we face at school. For example, can they handle a device as basic as a telephone? Its misuse is obvious. For example, that we put the phone down, go for a walk and let the other person continue talking.

We need inner strength for using a machine in a meaningful way. We need a kind of inner uprightness, an inner force to really be able to use them appropriately. I don't even want to mention the car! In this sense, it is a primitive machine and not very efficient. In fact, the car should not be allowed at all, because the degree of its use is a disaster. Most of the energy is not used for the journey, but is released into the environment, and to top this, it is a dangerous instrument; a rather dangerous machine for transporting our bodies from one place to another.

When you use a smartphone, you might think it is harmless. But it might not be that safe, because we also don't realise what this thing is doing to us; how it changes everyday life and how we can become addicted to it. Can I take my smartphone and say, now I'll consciously put it aside for a fortnight and do so?; calmly and without inner stress? Or are we no longer able to do that? Have we already become that addicted to this device?

How do I deal with these media? Or as a teacher, ask the question: What do I have to build up internally in order to use this technology meaningfully? This needs extensive training so that it is not introduced too early and so that the question is also asked: What do I have to develop inwardly so that I have the will to say: Stop! This far and no further!

A study was carried out on the behaviour of students at the computer screens, i.e. what they actually did during an online event was examined. These were students, and lecturers as well, and the interesting thing was, they all played a lot, and didn't use the device in the way it was supposed to be used. Instead, they worked a bit, then played another game or they worked and then sent an email and such things. It's no longer the most modern study, because today you wouldn't necessarily send emails. However, students have always been distracted, always. One wasn't consistently at work, but sometimes followed what was going on and sometimes meandered away. Where does the strength come from in this digital age, that I do what I set out to do and not let the machine steer me elsewhere? This is one of the essential questions that we ask ourselves at school: Can we develop something in young people when we talk about technology and digitalization? At what point do we even discuss media, so that the students themselves are able to recognize something in such a way that an array of perspectives open up?

A universal question today is: Do I have the peace of mind to look in many directions internally, when considering different perspectives, for example, in the evaluation of COVID-19? For instance, was there a single musician in any of the decision-making committees or were there mainly only virologists? Was there perhaps a young person involved in the deliberations who was only 15-years-old and would have also given an opinion on what he/she thought about things? It is important that an abundance of points of view are initially present, that the many contexts and points of view are gathered together in order to evaluate and order them, and from which an assessment can be made. What is important, what is less important? This ability is what we try to develop in Waldorf education: that our actions are inspired by such insights, and vice versa, that one gains new insights from one's actions, including learning from what one has done, or has been done previously. You do something and now you look at how it went. This is also a process of cognition: what are the consequences of what I have done? These are the most important processes that enable the use of digital mediums in a meaningful way.

Claus-Peter Röh, what would you say, if you had to take it from here?

Art in Education

Claus-Peter Röh: I do understand that a process should be stimulated in order for them to get involved in understanding technology, for observing new things and for exchanging ideas. At the same time, it is also about the appraisement of responsible action. So I have to combine several things into one. That's quite a challenge for a person who doesn't allow himself to be distracted and who has perhaps even learned during these Corona months that these devices are no longer just devices of leisure and amusement, but are necessities for work. As a class or lower school teacher, I often hear the question: What skills should I develop in the diversity of a class, and how do I address this topic within the framework of the whole school?

Here I must briefly talk about art in education, not just about the individual art subjects. If we extend the concept of art to the whole of education, then it becomes an 'art of education'. That is to say, a process in time which, even before I consciously get to know and practise things, a skill is being developed for later. Edwin Hübner[4] calls this 'indirect media education'. It already begins at birth! Strengths must be properly inwardly instilled: confidence, the joy of living, the experience of self-determination, of decisiveness and of getting involved in something—all this is indirect media education. This naturally leads to a conscious engagement in the middle school, because most pupils already have their own devices nowadays.

What role does art play in this process through time? Just two small examples from a school: a class-one child in Frankfurt arrives by underground and walks through the school corridors wearing a mask. In the classroom, they begin by running around. Everything happens quickly, one thing after the other, and it is always the inner being of the children that has to make all these transitions. While they are running around in the classroom, a bell suddenly rings. Everyone knows the signal: the lesson is about to begin. In a flash, everything changes and the lesson begins. The teacher speaks, welcomes the children and they speak the morning verse. Different rhythmical exercises are practised. She begins to tell a story about a boy and a girl, which leads them to what is new today. At the end of which, she makes a sign, which the children understand. They take out their beautiful writing books and open them. The teacher begins to write something on the blackboard. The children, with looks of curiosity and questioning, show their interest, clearly emanating from within. What's coming now? Today it is about a sound that consists of two sounds, E and I, but when together, they become EI, a sound, which is new for them. Some know it already, while others really examine it and rehearse the 'ei' sound. And while they are rehearsing it, they draw it, work with it in varying ways allowing it to come alive and then they finally show their work to others. It is a whole process, a teaching event, a single stream in an artistic sense. The children were able to get involved in the various steps of the whole process and work out of their own free willing. Then the children went out to break.

Class 12 follows, a different theme and artistically different. They are busy with literature, Juli Zeh's *Corpus Delicti,* which interests the young people. I'll just mention one pupil's comment, which makes it clear how we have arrived at the development of free thought formation. At a certain moment, she said, 'At this point, I see it differently. From my point of view, the control mechanism of this healthcare state, which at first seems to be completely concealed, had already been set up long before in the first chapters.' And then she gave her reasoning. So at this age, I can give free rein to my thoughts and enter into the art of conversation with others—educational artistic conversation, engaging with the different views of the others, voicing my own views and throwing new balls into the conversation with my 12th class peers.

In this sense, the artistic plays a crucial role in terms of bringing things together. Can we strengthen the children in such a way that later, they develop an inner decisiveness when viewing the whole?

That would be my contribution. Florian Osswald, what else could be done to ensure that the way we deal with technology remains human or at least comes very close to it?

What is the right atmosphere for learning?

Florian Osswald: It is a great art to deal with the realm of the in-between spaces in the school. We haven't even really looked at that yet and it is a very challenging situation today. In a situation such as we have at the moment, with pupils and teachers in front of screens or perhaps all wearing masks, how can we design lessons? One of the most important tasks, or actually the greatest responsibility of teachers, is to create the right atmosphere and space for an educational process to take place. And of course, this is not easy to achieve online.

Conversely, however, we can also say to ourselves: Now a new situation has arisen for schools, which had not previously been imagined, resulting from this Covid world in which we now live, raising many questions precisely with regard to educational space. Before, it was simply provided by the classroom, by the walls, which actually had a somewhat confining effect and in which not all pupils were completely comfortable. For instance, there are some pupils who would prefer to be outside in the garden and not in the classroom.

In 2020, we have had so many different experiences at school that would not have been allowed before in a Steiner or Waldorf school; they would have caused a revolution. In other words, it is a whole new world that now signifies school, and the question now arises: What kind of spaces do we actually need for this and where would school need to be located? Is it, for instance, only in a classroom? Up to now, schools also have gardens, a gym, handicraft rooms, and so on. They have different rooms. And now, we really have to ask ourselves the question with regard to educational space: Which spaces have the right atmosphere for these young people? Have we really posed the question radically enough to ourselves?

This is one of the questions we are dealing with. What kind of spaces would we have to design for and with our pupils in the future? Wouldn't there be other ways of locating a school, so that pupils don't always have to be physically there? That they don't always have to travel such long distances, but can perhaps also have things to create in their own environments and through this, also develop new thoughts and experiences in the process. That is the first thing.

The second thing that shapes the atmosphere to a great extent is the way time is shaped. I question the timetable of the school day, for instance. We have known for a long time that the timetable is not ideal. Yet we have not changed it! Again and again, we fall back into these old patterns and, of course, once the whole thing is up and running, it's up and running. I think we have to think about this question as well! For example, is the continual changing of subjects healthy? We have known for a long time that it is not good for all children. But we do it anyway because it's the easiest thing to organize. Couldn't it be organized differently in the future? Because a lot of things have now softened during the Corona era, the question has arisen more and more clearly about what the right learning atmosphere is for children and young people. How do we create an atmosphere in which the children feel comfortable, in which they enjoy learning, in which they enjoy discovering the world?

It's like a small child. If the child lacks a feeling of safety and security, then it cannot conquer the world around it. The child then retreats because he or she is afraid to go out into the world. For this exploration, the child needs a safe place to come back to. Already when the children are small, place plays a decisive role. How do we create that, an environment for children? I think this is something which we have a lot to contribute to, because we have had an infinite number of experiences in Waldorf educational institutions. However, we have to take the initiative and put it into practice.

It's a wonderful feeling knowing everything. You can read study upon study, one after the other, and everyone then declares: Change things! However, in school in general, nothing has changed. Only very few schools have radically changed things, on a trial basis. 'Doing good' would be the task of the future! It's nice to know truths, but it is important for schools to implement changes and learn new things from their experiences. That would be one approach.

A Suggestion for Video Conferencing or Online Teaching

So, Claus-Peter, what do you think is crucial for this situation? How can we support the children and young people even more, from your point of view?

Claus-Peter Röh: From my point of view, what is most important are soul spaces in which an inner response from the pupils to what they experience is possible. This is the primary basis for all learning—an atmosphere that provides the possibility for expressing one's own inner response to what is happening, or said in another way, that gives birth to individual responsibility. For what is responsibility actually? Let us not just look at technology, but also, for instance, at the earth's climate. Not only are many young people rising up in response, but they also feel personally co-responsible.

Can we create spaces in and out of the school in which inner and very personal responses can emerge, which lead to impulses for action? This has a lot to do with the artistic path I described: I experience something, I encounter things directly and create something in the process. In recent literature, it has been finally accepted that art in school is not just 'nice to have', a nice little extra, but is of integral importance in the whole learning process. Because what does art accomplish? It enables a slowing down of these processes, particularly facilitating a connection to and deepening of what is being worked on.

Secondly, art provides a sense of completion. Pupils themselves notice that something is missing, and their inner sense of responsibility at this level wants to take action to complete the whole. I often experience this very clearly during video conferences, and online lessons: this inner voice of responsibility says that it is not enough just to describe nice things, which might even reach the person over there. You have the feeling all the time that you need to compensate for something—which, by the way, is also exhausting—that is not there as it is when we are physically together. You feel that afterwards that you tried the whole time to compensate for something that was missing.

Would it be a bit more human in dealing with online events if one did not just speak, transmit content, consciously follow thoughts, but also enable a creative process for everyone? To illustrate: if each participant would draw sketches relating to what is being said? They all have large sheets of paper on their desks and create by hand what they hear on video. They create their own pictures of what they are receiving digitally and therewith also take ownership of their own image creating, their own understanding. My suggestion would be to deal with the current video and online educational situation in such ways as this.

What else would you say, as possible concluding remarks? What new possibilities could there be with this type of technology?

Florian Osswald: We tend to think that we learn in order to be able to use technology. However, we learn things in order to better use these technologies, so that we are able to find our way around in the world more conveniently or more quickly. This is one of the basic paradigms in which we live. I need, of course, to be strong-willed so that I can use technology in a way that I have determined and intended. It is therefore useful to study the diversity of technology.

One could also ask oneself, and this is always a question we ask ourselves as teachers: How can I design lessons that flow from the child itself? I want to create lessons out of what is present in the children or people now, out of the direct needs of the other. The question I pose to modern technology, to modern media would be: What and how can we learn through you? That we engage in media in such a way that it shows us its possibilities. That we don't just learn how to use media at school, but that it shows us how it has actually already changed the world and that we, because it exists, are able to offer new forms of teaching. This is a challenge, and at the moment we still have a lot to learn. It is important that we get to know these technical devices so well that we can design new kinds of lessons based on this knowledge. Of course, always with the underlying condition that we do whatever we do within an educational and meaningful context. Could it be that through the interaction with technology—which currently shapes our lives and world and rolls forwards and onwards and for which no one is to blame—that school life could take place differently?

Could we perhaps also carry out experiments, whether pleasant or not, that suddenly one day, let's say, the power is switched off? Thus we could only meet person to person, in the teachers' room, in the classroom or at home, so that suddenly we also rediscover the human being through a temporary renunciation of technology? I think this would be an exercise from which we could learn a lot, in order to develop an even greater inner sense for the place and task of technology in the present time. What really happens, when we meet each other, standing upright in direct conversation as human beings, without the distraction of technology? Perceiving the other person as a human being could become the new motto of the future.

Slightly edited dialogue lecture, 7 December 2020 at the Goetheanum.

1 Paul Klee, exhibition *Bewegte Bilder* [Moving Pictures], Bern 2016/17, notebook entry. Translation by C. Howard.
2 Rudolf Steiner, *The Bridge Between Universal Spirituality and the Physical Constitution of Man* (GA 202).
3 Remo Largo, *BAZ* 24.10.2018, *Kinder werden zu überangepassten Wesen*. ['Children are becoming over-adapted beings'. Development specialist and best-selling author Remo Largo doesn't think much of today's school system. An interview.]
4 Edwin Hübner, *Competency and media literacy*, January 2011, in Erziehungskunst, Waldorf Education Today. Accessed online on 25 June 2021, at:
   https://www.erziehungskunst.de/en/article/digital-self-defense/competency-and-media-literacy/

Christiane Haid

The Hidden Sun

Are art and culture really not 'system relevant'?

The corona image, the sun obscured by the shadow of the moon, with an impressive play of light forming along its black edges, was a symbol for me already in the first lockdown during which all things cultural: music, theatre, opera and museums, were shut down, just closed off. You may read literature or watch films and videos instead, however, art unfolds primarily in social space and is based on bodily experiences. If flight into the digital is the last resort in the given circumstances, then art is deprived of its effectiveness to a great extent. If art, whose effect reaches deep into our physical well-being, can only reach us from the surface of a screen, literally 'superficially', it breaks away from space and time, the two aspects essential to it. The art pieces are merely surrogates on a surface, at best a memory of what they actually are.

When the Art Museum Basel (*Kunstmuseum* Basel) records that the average time spent by internet visitors for its online tours is four minutes compared to visits to the museum, which otherwise last around two hours, the consequences of the loss are already visible in the ratio of the numbers alone. Neither an adventure nor an experience is available online for the forming of a relationship and human development. Thomas Fuchs formulates the necessity and significance of bodily presence of the human being in his latest book, *Verteidigung des Menschen* [In Defence of the Human Being] as follows:

> For what is in question today is what one could call—with unavoidable vagueness—the humanistic ideal of humanity. The human individual as a bodily or embodied being, as a free, self-determining and ultimately a constitutively social being connected to others, is at its centre. According to this understanding, people are therefore not mere spirits or monads of consciousness, but embodied, living beings. And people do not exist as singular entities, but only in common relational space.[1]

In recent months, I have read various articles in art categories about the lockdown of culture and its consequences. All the appeals to governments, particularly not to close down concert halls, theatres and museums, all of which had developed elaborate safety concepts, were unsuccessful and faded into nothingness. The closures are therefore all the more astonishing and remain questionable.

It is also thought-provoking that in government press releases opera, concert halls, theatres and other cultural institutions are ranked on a par with brothels, amusement parks and sporting events. This equation is not new and has been common practice for years, but the Corona crisis has now brought it clearly to the fore and shows that culture and art—and the humanities can also be counted among them—have become meaningless and seemingly dispensable. However, they are in fact necessary for our existence. They enable ways for people to develop inwardly and they lead humanity to self-responsibility and self-determination. Art and culture create free spaces in which the future of humanity can be shaped.

A Look into the Past—Art Heals

When I was 13-years-old, I attended the 7th class in a Waldorf school. This class was made up of pupils from all different kinds of schools and the social situation was extremely difficult. We had a brave and unconventional class teacher who, together with the music teacher, performed a prelude and fugue from Johann Sebastian Bach's *The Well-Tempered Clavier* and one of Novalis' *Hymns to the Night* every school day for a year. That's how the day began. This is perhaps amazing and unusual, some would even say, almost crazy, but this experience left a lasting impression on me. I read a lot of Novalis in my youth. He virtually kept me afloat during the difficult phases of my life. The class grew together through this artistic work during this extraordinary year. What was difficult and fragmented at the beginning, became an ever greater unified whole.

Later, when doing practice teaching for my training in a secondary school, and was engaged as a German teacher, I had a second incisive and ground-breaking experience of what art does. The pupils came from so-called educationally deprived backgrounds. I enjoyed teaching Goethe, Schiller and Hölderlin, for which my instructors sometimes gave disapproving glances. The opinion prevailed that such content could not be understood by the pupils at all. In one 8th class lesson, Schiller's ballad *The Veiled Image at Sais* was the topic. We started talking about why the young man, contrary to the priest's command, unveils the veiled statue one night. He is terrified by what he sees and dies a short time later. I then asked the pupils how it was that the young man died as he had been searching for the truth and had been told that the statue was the truth. Then a slender boy stood up, and very excitedly shouted into the room in a clear voice: 'Yes, because he saw *his* whole truth.' These are moments in which it becomes impressively clear that extraordinary things can happen through art. For a moment, an insight of unheard-of depth became possible for the pupil, which shook and thrilled me with regard to what culture and art can unleash in people.

Unfolding

Adalbert Stifter's great novel, *Indian Summer* (1857) illustrates another feature of the effectiveness of art. Stifter wrote this novel as a counterbalance against emerging materialism. It is a prime example of the experience of deceleration. It is a thick book of around 800 pages that opens up an incredible world. A world that breathes tranquillity, that shows an admirable care of perception of the sensual world and that, one could say, opens up an immeasurable space for development. With a love of the smallest details, the life of Heinrich Drendorf is traced, as he pursues his own interests without any outside influence, coming more and more into his own. He has been given a certain amount of money by his father, with which he must manage, otherwise there are no expectations placed on him.

He decides to study the natural sciences, and his father is resented by those around him, who have no understanding for such a generous and free education. On his journey, he meets the Baron of Risach and soon becomes a regular and welcome guest in his house.

Old Risach and young Heinrich have a conversation that exemplifies what the perception of art can develop into. After Heinrich has been a guest of the baron for several years, the baron advises him to do nothing for a summer; not to rush after a scientific problem or pursue some project, but just to let himself drift; not to take on anything at all and see what happens. And so, what does he do? Heinrich takes the advice. He begins to draw, nature at first. He roams around the country estate, examines things closely, admires the beauty of the plants. One day, his attention turns to the art collection in his host friend's house. He takes a closer look at it and discovers a special sculpture in the house and then falls in love with it:

> 'Why didn't you tell me,' I continued, 'that the statue standing on your marble staircase is so beautiful?'
> 'Who told you so?' he asked.
> 'I saw it myself,' I answered.
> 'Well, now you know it with more surety and believe with greater certainty,' he replied, 'than if someone had given you an opinion about it.'
> 'I do believe that the statue is very beautiful,' I replied, trying to improve myself.
> 'I am sharing with you the belief that the work is of great significance,' he said.
> 'And why have you never spoken to me about it before?' I asked.
> 'Because I thought that you would look at it yourself after a certain time and consider it beautiful,' he replied.
> 'If you had told me sooner, I would have known sooner,' I replied.
> 'To tell someone that something is beautiful,' he replied, 'is not giving someone ownership of the beauty. In most cases, the other must merely believe it. Certainly, therefore, one thereby deprives that person of the ownership of beauty as he might have come to it of his own accord anyway. I took this for granted with you, and that is why I was happy to wait for you.'
> 'But what must you have thought about me all this time that I could have seen the statue, and still you kept silent about it?' I asked.
> 'I knew you to be truthful,' he said, 'and I held you in higher esteem than those who speak of the work without conviction, or than those who praise the work because they have heard others praise it.' [2]

This dialogue is paradigmatic of an entirely hierarchy-free relationship between the conversing partners; one which trusts in the capacity for development of the other. The older person, who could have imparted his knowledge, waits until the younger person discovers beauty out of his own perception and sensitivity. The host renounces passing on knowledge because he values Heinrich's independently-arrived-at judgement. Heinrich should not merely 'trust' that the statue is beautiful based on the judgement of his host and friend, but 'perceive' its beauty for himself. This is a subtle but crucial difference between accepting a judgement on authority and having the opportunity to form it for oneself. The latter possibility goes much deeper and is more sustainably connected to who I am. As a result, it also forms a corresponding competence, furthering judgement-making abilities from within. The attitude of the host expresses the highest respect for the freedom and autonomy of the other. He should not just adopt the thinking of others, but let his own abilities develop individually and appropriately when it is time for them to do so. In this way, Heinrich discovers the beautiful entirely out of himself, out of freedom.

I feel this to be an outstanding passage, in which the breath of freedom flows, freeing Heinrich's personality to unfold in an invaluable way.

The Science of Freedom

From this point of view, it is no coincidence that in one of Rudolf Steiner's early works, *The Theory of Knowledge Implicit in Geothe's Word Conception*[3], which he wrote at the end of his time in Weimar as Goethe's editor, he describes spiritual science as the 'science of freedom'. In this work, Steiner summarized philosophically what Goethe lived cognitively, but could not and would not describe systematically. The book is inspired by Goethe's method of cognition, but forms a self-contained philosophical approach. One can see this early writing of Steiner's as a paradigmatic sketch for the forming of a future science. The core idea of the work is drawn from Goethe's particular way of looking at the world. In several places, Goethe formulates that the methodology of science must result from the specificity of the particular object being investigated. Thus, there is a difference that will affect the method of cognition, whether I'm investigating how a falling stone behaves, or whether I'm observing the growth of a plant, or whether I'm contemplating a work of art. These are each three different layers of reality that require different forms of cognition. The stone belongs to inorganic nature and demands a method appropriate for knowledge of the physical/material. In this layer of reality, cause-and-effect relationships that can be physically grasped are considered. Their details are openly accessible to our understanding and our thinking can grasp all their facets. In the phenomena of organic nature, we recognize the idea of 'typus'. It underlies the manifold forms that shape the individual phenomena of nature as a law which determines them, but which does not appear visible from the outside. The type-idea remains in the 'ideal' sphere, as it were, and is supersensibly perceptible and conceptually imaginable. In this sense, natural science is a science, as Rudolf Steiner puts it, in which the 'completion of the work of creation' comes into being only in the thinking human being. In the natural sciences, both inorganic and organic, the human being grasps the underlying law from the diversity of the individual phenomena through thinking. This way of methodically dealing with world phenomena can be described as nature's confrontation with itself, reflected in the consciousness of the human being. In this sense, thinking is the last link in a sequence of processes that nature brings forth.

Art, as well as the spiritual sciences, have completely different qualities than the natural sciences. Human consciousness is directed towards content and phenomena that the human spirit itself creates, insofar as the subject and content are incomparably closer to it. In philosophy, in historical events, in cultural phenomena, in literature, in works of art, we are looking at the achievements of the human spirit, at what it has brought forth from out of itself. In this sense, all human creation contains within itself the law that causes it. Hence, in *The Theory of Knowledge Implicit in Goethe's World Conception* Rudolf Steiner says:

> The spiritual is grasped by the spirit. Here, reality already has within itself the ideal element, the lawfulness, that otherwise emerges only in spiritual apprehension. That which in the natural sciences is only the product of reflection about the objects is here innate in them. Science plays a different role here. The essential being would already be in the object even without the work of science. It is human deeds, creations, ideas with which we have to do here. It is the individual human being's coming to terms with him[- or her] self and with their belonging to the human species.[4]

It is precisely the latter, 'the individual human being's coming to terms with him[- or her] self and with belonging to the human species[5]' that is central to the meaning and task of the entire sphere of culture, art, philosophy and history. For in these manifestations, the realities that create identity, meaning and self-knowledge exist, which are indispensable for understanding one's own humanity. For instance, the passage quoted above from Adalbert Stifter's *Indian Summer* is able to stimulate thoughts about the development and self-determination of the human being without any orientation towards action; in this artistic creation it remains a free-flowing suggestion, a possibility for identification, which the respective reader may or may not respond to. Whether it will have biographical relevance is up to each individual person.

In the mirror of the works of other people, that is, their spiritual creations, the human being is able to see itself and the world in a completely new way. It is an astonishing and at the same time inspiring fact that we have the ability, because we ourselves have spiritual faculties—the artistic manifestation is, after all, produced by a spiritual being—to perceive spiritual things. Of course there are degrees, and it is a question of ability and education how much each person can understand or perceive, but basically spiritual perception is given to us all as potentiality. We perceive something in this sphere that cannot be perceived in the other kingdoms of nature because they were not created by us. We encounter the kingdoms of nature as creations already in existence and can recognize them by means of our thinking, because we too are part of this creation. If we now consider the creations of the human spirit itself, we are the creators, and by engaging with the creations brought forth by human beings, be it through recognition or enjoyment of them, we enter into a relationship with the manifestations of the human spirit and experience human creation and activity, and thus ourselves as human beings.

What is the human being?

From 4 July to 8 November, an exhibition 'Into the Unknown: Art in the Times of Coronavirus' took place at the Goetheanum. This exhibition was an inspiration born out of necessity. As we were unable to do anything for the public due to the lockdown, the idea arose in the Visual Arts Section to invite our artist colleagues to make a work created during the Corona period available for an exhibition at the Goetheanum. This call was answered by 52 artists from Europe and overseas who sent a total of 82 works[6] to the Goetheanum. Each artist was also represented with short written statements.

The work pictured here is entitled 'König' (King). It is by Barbara Schnetzler, who is part of the management team of the Visual Arts Section and curator of the exhibition. In the process of working on the sculpture 'König', she tried to explore boundaries. How much can be taken away from a face until it is no longer there, so to speak? At present, we are all partly experiencing a strong reduction in various areas of life: we cannot travel, are strongly restricted to the local and are determined by state regulations. The pandemic has shown the vulnerability of human beings with regard to their health. Barbara Schnetzler has tried to express this vulnerability in her sculpture. The title 'König' evokes certain associations, such as the royal, rulership and the sublime. The face with its over-long neck is built on an iron rod, which appears as a support that provides stability, in the overall impression of decay that this sculpture has as well. If you look at the eyes, the empty sockets immediately convey the impression that there was once life there, but which is no longer there. For Barbara Schnetzler, the central theme in her work is the tension between the beautiful and the ugly. To show the beautiful in the ugly or the ugly in the beautiful is something she seeks in her art. In addition, there is an aspect of uprightness, the uprightness that is visible in the rod holding the head, that gives the work a vertical strength and dynamic. It is a kind of inner I-gesture. When I looked at this sculpture more intensively, I thought of what Meister Eckhard said: 'A king who does not know that he is a king is no king!' Considering these words together with the statement made by the sculpture itself, the question can be asked: Is the royal still there or has too much of the physical substance of the sculpture been removed? What is the relationship between the determination of being a king and the appearance of the sculpture? One may inquire about the relationship of human beings to their unique characteristics as spiritual beings, and at the same time, to their own physicality. What constitutes the spiritual, what constitutes the inner orientation of the human being, regardless of whether he or she is a physical earth being? How is the connection between material and spiritual existence to be understood and what kind of life develops if it is limited to 'the bare essentials', to the pure preservation of the biological?

What do I mean when I speak of the human being? Are we a biochemical mechanism that drives brain and spirit, that understands feelings as biochemical reactions? Or is the human being a body, soul and spiritual being that finds itself at home in a shared world, as a counterpart of fellow human beings, in nature and the cosmos? This question is formulated here with good reason. For in the Section for the Literary Arts and Humanities, we have, for some time, also been dealing with the ideas of transhumanism, among others.

It is interesting that Ray Kurzweil, chief engineer at Google, makes the prophetic statement in his book, *Homo S@piens* (1999)—prophetic statements are one of the peculiarities of transhumanists—that before the end of the next century humanity will have lost its position as the most intelligent and efficient beings on earth. However, he qualifies this prediction by pointing out that the realization of this depends on what we mean by the term 'human being'. For Kurzweil, the question of what the human being is, how the human being is conceptualized, is the most pressing political and philosophical question of the present. I believe that against the backdrop of the events of the pandemic, this question becomes even more acute when we witness what is happening to art and culture, that is, to the whole spiritual sphere.

Yuval Noah Harari, a currently widely read author, takes Kurzweil's thesis further into a bioscientific direction. In his book, *Homo Deus* he says: 'People will no longer regard themselves as autonomous beings who lead lives according to their own desires, but will become more and more a collection of biochemical mechanisms that are constantly monitored and directed by a neural network of electronic algorithms.'[7] And a little further he says: 'Free will exists only in the imaginary stories we humans have invented.'[8] This is the perspective that emerges from the future visionary direction inspired by transhumanism, by biotechnology, by the life sciences. I think we need to really ask ourselves, is this still only a prognosis or is it slowly becoming a reality?

Self-determination and Self-responsibility

What could the spiritual sciences, culture and art contribute to the quality and tasks related to these questions? As already mentioned above, in *The Theory of Knowledge Implicit in Goethe's World Conception*, Rudolf Steiner calls the spiritual sciences the science of freedom. He distinguishes the domain of the spirit from the other two, the organic and inorganic. The following quotation makes clear what dangers lie in an absolutism that makes one domain the only valid one for interpretation, and which thereby deems the other domains as extraneous for human beings:

> Man should not, like a thing of inorganic nature, work upon another being in accordance with outer norms, in accordance with a lawfulness governing him; he should also not be merely the individual form of a general typus; rather

he himself should set for himself the purpose, the goal of his existence, of his activity. If his actions are the results of laws, then these laws must be such that he gives them to himself. What he is in himself, what he is among his own kind, within the state and in history, this he should not be through external determining factors. He must be this through himself. How he fits himself into the structure of the world depends upon him. He must find the point where he can participate in the workings of the world.[9]

It is in the nature of the human being that the individual determines his or her destiny as a free being. This does not mean excessive egoism, nor does it mean ignoring the human being as a social being. It is about a basic principle of inner orientation, how life is to be shaped. The training of this inner determination is precisely one of the goals of human development. It could also be expressed in a more modern way as the ability to creatively shape one's *self* and the world. Self-determination and self-responsibility then also lead to social shaping, something which fulfils the human being, and lies potentially, as a basic idea, in the arts, in the areas of culture and in the spiritual sciences. It is precisely this potentiality that, in my view, is currently being increasingly restricted and questioned.

In Conversation with the Unknown

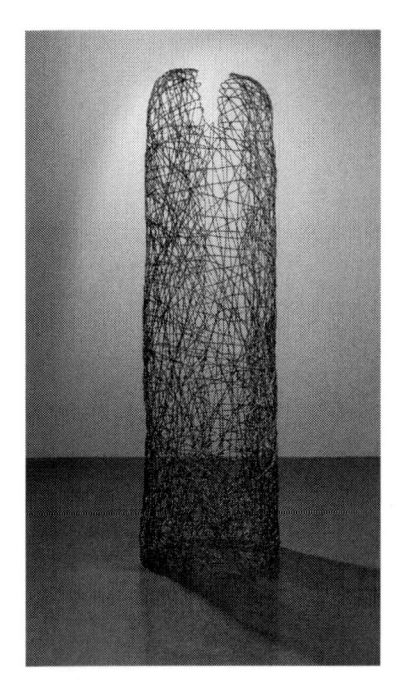

The iron sculptor, known as FEROSE, created this work[10] during the Corona period. FEROSE told me she had welded the sculpture partly during the day and partly at night, in a kind of creative thrust.

When FEROSE creates a work of art, she initially has an idea in the form of a gesture that she experiences within herself. In the creation of the artwork, however, these preliminary drafts disappear almost completely. Iron, the material primarily used by FEROSE, is very hard and at the same time, when warmed or heated, becomes a pliable material. Our culture is inconceivable without iron. We live in an iron age. Our machines, our means of transport, and so on surround us with enormous amounts of iron. At the same time, the iron in our blood is what makes us strong. If we are iron deficient, we are not able to work.

The iron sculpture shows a kind of shell of a human figure made of a fine mesh of iron rods welded together in various ways. The shell is without a head and one could also think that the head is enclosed, and extends, as it were, from the shoulders to the ground. No bodily features stand out; the figure can be experienced from the shoulders down as if in a kind of wrapping or garment. Standing in front of the sculpture, one gets the impression of great fragility and tenderness. At the same time, as a viewer, you can step out of yourself, immerse yourself directly in the figure, experience yourself in its enclosure and breathe with it. This creates a dichotomy of sensation: of being supported and of being held in captivity.

FEROSE works with an open welding torch, so directly with fire. She told me that she works so that the iron, which is grey to begin with, virtually goes through all the colours of the colour spectrum. She designs some of her works in such a way that she also dyes them herself. For example, art pieces with a strong blue shimmer are created because the processing with fire is designed in such a way that this blue emerges. FEROSE writes in her short text about the depicted work:

> In the last few years, I have worked in isolation and concentration in my studio. Corona intensified the outer calm, and broadened my view on the subject of enclosure. I have had to muster strength to transmute the restlessness, injustices and the questions 'in the air' in order to connect with people in spiritual communion.[11]

FEROSE sees art as a transition into a further, one could perhaps even say higher or next stage, of social sculpturing. She is inspired by Beuys and his statement: 'Every human being is an artist.' In working artistically, in dealing with uncertainty that is so strongly experienced at the moment, lies the possibility of entering into something that not only comes from within oneself, but arises from entering into conversation. Something that slowly develops into something real, something that you cannot create on your own. This is the path towards a new form of the social.

Overcoming Fear and the Fear of Death

Art stirs us at our innermost core. This is particularly evident in the relationship between death and art. Death was the dominant theme last year: negative, to be avoided and filling everyone with fear. The term 'death' is one of the most frequently used words in the media, after the word 'virus'. The driving force behind this is to escalate the fear of death, which is increasing not least due to media reporting. Our relationship to death is the central question par excellence at a time shaped by pure materialism. The fear of the end of bodily-material existence and the threat of emptiness and nothingness are the root causes.

The visual arts, philosophy and literature offer an unbelievable treasure of experiences, for ways of dealing with and even for integrating the experience of death, up to and including completely fresh perspectives on its nature and meaning for human beings. For the artistic process, death even proves to be a necessary and impelling borderline experience where creativity unfolds in the first place.

There is another passage in the book, *Homo Deus* by Harari, already quoted, that shook me up. It says: 'As soon as people believe (whether for good reason or not) that they have a good chance of escaping death, the desire for life will make them no longer want to pull the rickety wagon of art, ideology and religion; they will plunge ahead like an avalanche.' [12] This statement can get under one's skin against the background of what we are currently experiencing. Death, as Harari describes elsewhere in the book, is now only a technical problem, given the prospects offered by the life sciences combined with biotechnology. The transhumanists dream that death is surmountable. But these visions of the future are entirely fixated on the material level, from substances that one ingests to downloading consciousness of a brain to a machine, in order that the human being is no longer subject to old age, illness and death. Here, a radical abolition of all transcendence is pursued in favour of total predictability. The 'brave new world' that would thus come into being would be deprived of its future, because the future arises from the uncertain, the not yet predictable, from the surprising happenings in a dialogue.

The Spanish author, filmmaker and temporary Minister of Culture of Spain, Jorge Semprun (1923-2011), had to experience first-hand, the worst of human catastrophes to date. He was imprisoned in Buchenwald concentration camp, near Weimar in Germany, during the Nazi era.

At the age of 20, he was arrested as a member of the Resistance in France and was then deported to Buchenwald. He describes his concentration camp experiences in several autobiographically inspired novels. He could not start writing immediately after his return; it took 16 years before Semprun was able to transform his experiences into art without losing consciousness or suffering from the worst kinds of anxiety. Until then, he had decided on conscious amnesia. The opportunity to reflect on his experiences in writing nevertheless arose quite unexpectedly. One evening, a host with whom he had gone into hiding, an agent against Franco in Spain, recounted his experiences in the German concentration camp, Mauthausen. The narrative poured over his listener like a structureless stream of 'unbridled accounting', almost robbing him of his composure and consciousness. Semprun did not tell his host that he had experienced the camp himself, but suffered through the narrative to the end. Subsequently however, this very 'unbridled untamed account' gave him the impetus to begin writing after 16 years of amnesia. At first, approximations of the camp took place in his imaginings; the imprisonment and the journey there are described in his first novel, *The Long Voyage Journey* (1981). The final step, however, of being able to penetrate the interior of the camp from memory, only became possible when Jorge Semprun himself travelled to Buchenwald in the early 1990s and physically walked through the place once again. He describes this in *Literature or Life* (1997), one of his last novels about these experiences. There, the protagonist, who saw himself in the reflection of the gazes of the American liberating soldiers, tells of how death is no longer the perspective in which he lives, but rather that he now experiences moving further away from death with each day of his life. Now he has passed through death, death is behind him, as it were. He has become a 'revenant', which he sums up with the words: 'I have nothing but my death.'

In reflection of the concentration camp experience, the question of how to represent and communicate these experiences comes up again and again. Authors such as Imre Kertesz, Primo Levi and Jorge Semprun have asked themselves this question in their accounts of their camp experiences, and in so doing speak of a sense of duty to convey and especially to bear witness to it for posterity. It is clear that merely documenting their experiences remains insufficient, because this hardly allows for the scope of the events and for the understanding of them to become concrete. So it is an existential question of how the experiences can be captured in such a way that they are not just a documentation of horror. This question is related to the essence and task of the artistic process—which is the reason I raise it here. It is about the question of what can be achieved through art. What is that special quality that arises through art which goes beyond a mere documentation of events, both for the author and the reader? I would like to draw attention to two questions. The first concerns the substance of what is being narrated. How can the author convey the substance of his narrative, which is so vastly different in its subject matter and essence than anything else normally experienced in life? How can it be grasped and shaped?

The second concerns the intensity of the material, which Imre Kertesz also talks about. The intensity refers to the nature of the experiences. They are so extraordinary and existentially profound that basically no one can comprehend them who has not known them from their own experiences. Imre Kertesz was once asked by an interviewer at the Basel Book Fair, which I attended myself, what he meant by 'happiness in the camp' in one of his works. She found this statement completely incomprehensible in view of the camp situation. Kertesz replied that he did not mean what we would commonly call happiness, but referred to a special intensity of existence that develops due to this extraordinary life situation. This then leads to another question, repeatedly reflected on by writers with concentration camp experience, of how the writer can recall such experiences at all, without becoming completely overwhelmed by them. The authors though feel the need to put their testimony into words, to make what happened accessible to the world and to recapture that life in writing. The process of putting experiences into words is sometimes an incredibly painful process for them, but it is also something that gives them a kind of new life. In *Literature or Life*, the first person narrator reflects on this:

> Yet I am overcome by a doubt regarding the possibility of being able to narrate. Not that what I have experienced cannot be described. It was unbearable, which is something quite different, as one may easily understand. Something else affects its substance, not the form of the account. Not its structure, but its intensity. Only those who know how to transform their testimony into a work of art, allow space for creation, will penetrate this substance and make the intensity transparent. Or, on re-creating: Only the

artistry of an unbridled account can transmit the truth of the testimony, and then only partially.[13]

One can see in the writing process—both Kertesz and Semprun report that in order to help them get down to their work, they made broad studies of the archives before beginning to write—that the process of re-creation itself expresses something that gives this monstrosity of human experience and historical events something that has transformative power, something that helps to transform this death, to transform it into a new form of existence. Art is the transforming force that raises what has happened above mere documentation and leads it into another layer of existence.

Dialogue with the Unknown

In this way, art enables new and different approaches to life that contain transformational power and which do not exist in any other area of life. I have already indicated that in the artistic process, the intention, the goal, is not essential for the realization of a work of art, as it is in everyday life or in other professional fields. One could even say, to the contrary, because if the artist were only to implement the ideas he or she carries around with him or her, then no art would come into being at all. We would have what is misleadingly called art today, products of Artificial Intelligence. What the artist brings forth is a creation of something that is not only determined by his or her intentions, but is a kind of dialogue with an unknown, a something that cannot be dispensed with. At the beginning of the artistic process, one sets something, a kind of impulse as it were, a first step. As a painter, for example, you put a certain colour on the canvas—and then the question is, does an answer arise from this uncertainty? The answer comes or it doesn't, then you have to wait. We know from Rainer Maria Rilke that he had to wait ten years for the completion of his 'Duino Elegies' and that before that, he went through an extremely dramatic period of crises and creative barrenness. The unknown cannot be forced, it is not accessible and appears at an unexpected, unpredictable or incalculable time.

When you have a look at artists' biographies, they are often biographies that have to do with tremendous borderline experiences, lives lived at the threshold, on the precipice of death, or accompanied by serious illnesses. Being an artist is a completely different form of existence that has precisely these borderline and threshold conditions, death experiences in life, as a basic condition for anything to come into being at all. In this respect, borderline experiences constitute the basic conditions for art and culture. At the same time, they give us the strength to endure the primal experiences of humanity, illness, old age and death, to grow through them and to mature inwardly.

A picture entitled *Tanz (Dance)*, by the Basel artist Roland Lardon Nordal, illustrates this. Roland Lardon Nordal is an art and crafts teacher at various Steiner schools in the Basel area and is also active in adult education. The departure point of this painting work is the *Basler Totentanz* (Basel's Dance of Death). These are motifs and pictures that were created as a result of the plague in the Middle Ages; the plague that seized everyone regardless of class and status and did not even spare the rich. This gave rise to many pictorial works, for instance, by Hans Holbein, among others. Roland Lardon Nordal has also engaged with them.

The basic colours of the paintings are a luminous light blue and gold. The theme of 'dance' seems a kind of veil over the deeper meanings that are connected with the paintings and their theme. Roland Lardon has dedicated many paintings to this theme. If we look at his painting, on the right-hand side, one can see a structure literally protruding from the picture. It is a plastic form, a figure imposed on the flat picture. He calls it a sheep. On closer inspection, I immediately had the idea that it must be the Lamb of God. He then told me that he has studied the Old and New Testaments a lot and that he had carved 70 lambs, one a day, for 70 days, during the Corona period. Some of them turned golden. The picture we see here shows the Lamb of God as a sculptural work, standing on a black pedestal and watching the dance of death. The lamb thus looks, as it were, to the surface of the picture which portrays the dance of death and is witness to the great numbers of death. If we consider the significance of the Lamb of God, it is the image of Christ, the One who overcame death. The image thus shows, as it were, the dance of death in the presence of the One who overcame death. The lamb is watching the struggle the people are engaged in, in their confrontation with death. However, it is not an observation as such, but a presence and a participation in this confrontation with death. The figure of Christ can be a help for people in a non-denominational sense to think further about the phenomenon and reality of death beyond a purely earthly thought of the material end of the body. The overcoming of death as a spiritual process that transforms and transcends the earthly was exemplified by Christ for humanity, as a potentiality. Understood in this way, the fact and reality of transcendence is not only an inner-religious and denominational process that one can merely 'believe' in the sense of Stifter, but if we add Rudolf Steiner's descriptions of the human being as a spiritual being, this sacrifice for humanity can be understood as a transcending process for all human beings.

However, if the image of the human being is only placed on a purely material level, then art and any consideration that transcends the purely material have no *raison d'être*. This is why the approach to knowledge that I mentioned at the beginning is of central importance. Building on Goethe's phenomenological view of nature, Rudolf Steiner described three layers of cognition. The question is answered as to which form of thinking follows from the subject matter itself, or which differentiated area of the reality is being considered, in the sense of *The Theory of Knowledge Implicit in Goethe's World Conception*: firstly, inorganic nature, with its cause-and-effect relationship; secondly, organic nature with the idea of 'typus' that shapes it; and thirdly, the realm of the spirit, with its creations emanating from individual spiritual human beings. In the sense of this Goethean method of cognition, if I apply forms of thinking that are appropriate only to knowledge of inorganic nature to the sphere of the human being, to social or cultural life, then a decisive part of reality fades away. One could even say that the human element is rendered invalid and is eliminated in favour of a purely mechanical-materialistic outlook. This results in wrong judgements being made and ultimately questionable actions. If we are to look from the perspective of the human being, then the question arises as to what 'nourishment' the human being requires in order to be able to live out his or her needs as a spiritual being? As a being of freedom, we human beings are dependent on our soul and spiritual existence being shaped and nourished. If we are only given 'inorganic' matter, we wither away and lose our purpose.

I will conclude with a review of a quotation from *The Theory of Knowledge Implicit in Goethe's World Conception*. It comes from the chapter 'Human Spiritual Activity' ('Human Freedom') and in its condensed form is like a summary, one could say a kind of meditation, on the relationship to the ground of the world, the entire world context, to human thinking. How we conceive thinking to be—whether purely materialistic and inorganic, or whether appropriate only to the organic, or especially whether appropriate to the realm of the spiritual with its existential significance for human cognition and art—is what our future depends on.

The quotation opens the horizon with the thought that the human being does not live as a being that exists separately from the world, but that the ground of the world lives itself out in human thinking. And thus there is no fundamental separation between human beings and the ground of the world.

> The ground of the world has poured itself completely out into the world; it has not withdrawn from the world in order to guide it from outside; it drives the world from inside; it has not withheld itself from the world. The highest form in which it arises within the reality of ordinary life is thinking and, along with thinking, the human personality. If, therefore, the ground of the world has goals, they are identical to the goals that the human being sets for himself in living and in what he does. It is not by searching out this or that commandment of the guiding powers of the world, that he acts in accordance with its intentions, but rather through acting in accordance with his own insights.[14]

Art and culture are fed by the uncertain, the unknown. In other words, their outcomes are not calculable. If one takes the perspective of the above quotation seriously, it becomes clear that the unknown is only apparently unknown. In art, culture and the humanities, it becomes effective and visible in the world through the human being's capacity for intuition and artistic ability, and raises human beings above the purely material plane, which is only one side of his being. The night and the breathing of the cosmos are sources of sustenance for us.

When it is said that art, culture and the humanities are not 'system-relevant', this is true in a different sense than the one that is usually meant. The spirit cannot be part of a 'system' at all; it is all-encompassing, cosmic and human, all at the same time. This makes it clear why we need the perspectives and horizons of art and culture more than ever now—they constitute the human in the human being—they are like the sun whose rays sustain life, warm it and make it healthy.

1. Thomas Fuchs: *Verteidigung des Menschen*. Grundfragen einer verkörperten Anthropologie, Berlin 2020, P. 8. [In Defence of the Human Being. Basic questions of an embodied anthropology.] Quote freely translated by C. Howard.
2. Adalbert Stifter: *Indian Summer*, translated by W.W. Frye, Published by Peter Lang Publishing. First published 1857. Dialogue freely translated by C. Howard.
3. Rudolf Steiner: *The Theory of Knowledge Implicit in Goethe's World Conception*, (GA 2).
4. Rudolf Steiner: *The Theory of Knowledge Implicit in Goethe's World Conception*, 'F. The Humanities, 17. Introduction: Spirit and Nature'. (GA 2). Quote translation slightly altered by C. Howard.
5. The word used in German is '*Geschlecht*', which can also mean gender, but the biological gender is not meant here, but the human species.
6. The exhibition is documented and published in the journal *STIL. Goetheanismus in Kunst und Wissenschaft*, Michaeli 2020, available from the Verlag am Goetheanum, the publisher at the Goetheanum. German only.
7. Yuval Noah Harari, *Homo Deus: A Brief History of Tomorrow*, Harper, Feb 2017. Quote freely translation by C. Howard.
8. Ibid.
9. Rudolf Steiner: *The Theory of Knowledge Implicit in Goethe's World Conception*, 'F. The Humanities, 17. Introduction: Spirit and Nature'. (GA 2)
10. Her works have no title.
11. See the journal *STIL. Goetheanismus in Kunst und Wissenschaft*, Michaeli 2020. German only.
12. Yuval Noah Harari, *Homo Deus: A Brief History of Tomorrow*, Harper, Feb 2017. Quote freely translation by C. Howard.
13. Jorge Semprun, *Literature or Life*, 1998, Penguin Books. Quote freely translated by C. Howard.
14. Rudolf Steiner: *The Theory of Knowledge Implicit in Goethe's World Conception*, 'F. The Humanities, 19. Human Spiritual Activity'. (GA 2)

List of illustrations

Barbara Schnetzler: 'King' (*König*), clay, shellac, iron, 69 × 24 cm, 2020, photo: Malina Haid.

Ferosa: Untitled, iron, 186 × 50 cm, 2020, © 2021, ProLitteris, Zurich. Photo: Renè Lamb.

Roland Lardon Nordal: 'Dance' (*Tanz*), tempera with sheep, 2020, Photo: Malina Haid.

Matthias Girke

What effects do inner work and meditation have on the healing powers of the human being?

Effective protection against infection is to avoid exposure. This also applies to the COVID-19 disease caused by the SARS-CoV-2 virus and its various mutations. All government measures are aimed at reducing the risk of exposure. However, many people are also concerned with the question: What can I do inwardly to develop and promote a strengthening of resilience and salutogenetic forces that can exert an influence on my own disposition to disease?

In infectious diseases, there is a simple equation that has been valid for a long time, which is a trinity:

Exposure and Disposition Determine the Infection

Several approaches for dealing with the current pandemic follow from this. One, is avoiding exposure to SARS-CoV-2, protecting at-risk groups therefore reduces the risk of exposure. Two, is through the strengthening of resilience and salutogenetic forces, which influence disposition. There's a lot that can be done to help in terms of lifestyle (diet, exercise, sufficient sleep), but also through inner development and meditation. Three, a proper medical approach is needed for the infection itself, which for example, also sees the initial febrile, inflammatory reaction as the immunological capacity of the person to overcome the infection, and then to direct it in an appropriate therapeutic manner. Currently, the therapeutic significance of fever is well documented in animal experiments as well as in humans. The pathological 'hyperinflammation', which the organism itself can no longer control, requiring a different kind of therapy, must be distinguished from the fever response.

What Influences Determine a Person's Disposition to Disease?

With regard to COVID-19, we know the risk groups that need to be protected, because they are particularly at risk due to *somatic* factors that influence the course of the disease. Chronic diseases, such as cardiovascular disease, hypertension, diabetes mellitus and COPD (chronic obstructive pulmonary disease), are risk factors for severe disease progression. They particularly affect the respiratory tract and the cardiovascular system, and above all, the organs of the middle sphere of the human being, the rhythmic organization.

There are also *soul* factors, which Goethe pointed out almost 220 years ago. He himself came into contact with 'putrid fever' that was going around in connection with the Napoleonic campaigns. We know it today as rickettsioses, which is transmitted by fleas and leads to typhus. At that time, he drew the conclusion for himself and dictated it to his 'secretary', Johann Peter Eckermann, that an inner attitude, a moral resolve that one carries within oneself, has a positive effect on the disposition:

> He really [Napoleon] visited those sick of the plague, and, indeed, in order to prove that the man who could vanquish fear could vanquish the plague also. And he was right! I can instance a fact from my own life, when I was inevitably exposed to infection from a putrid fever, and warded off the disease merely by force of will. It is incredible what power the moral will has in such cases. It penetrates, as it were, the body; and puts it into a state of activity which repels all hurtful influences. Fear, on the other hand, is a state of indolent weakness and susceptibility, which makes it easy for every foe to take possession of us.[1]

Today, we are not only dealing with the pandemic of SARS-CoV-2, but also with a pandemic of fear and anxiety, which has a negative impact on our disposition, and on the immunological response potential to the disease.

Therapeutic Effectiveness of Meditation

Meditation helps the human being to open the soul up to the spiritual, to come 'from a life in thought to a life in spiritual beingness'. This 'upward-directed' quality of meditation, which leads one beyond oneself, consequently connects one to the spiritual being of the world.

In the Latin word meditation, the concept of the centre 'medium' also resonates. Many people have the need to find their centre because they feel that they have lost their inner balance, that is, they have 'left their centre out'. Anything that brings our feelings, our soul, into too much tension and oriented towards consciousness and thus towards the nervous-sensory system, has a negative effect on the immune system. The immunosuppressive effects of stress and emotional strain have been known for a long time.[2,3,4] Thus, there is a strong longing of many people to come 'back to their centre' through inner development, through inner work.

The word meditation also contains the Latin word '*mederi*'—to heal. Meditation is therefore also connected with healing and with health.[5,6] Especially in the context of the COVID pandemic, this opens up perspectives for developing healing powers through inner work. We now know about the healing power of spiritual activity. Jon Kabat Zinn and many others have researched the question of 'mindfulness'. How does meditative work affect the organism? What does it do for the nervous-sensory system and the cardiovascular system? And finally: What are its physiological effects on the processes of the metabolic-limb system, as well as on inflammation, immunity and regeneration?

Meditation and the Nervous System

Meditation has different effects on the nervous system. Numerous studies show how the grey and white matter of the brain is affected by meditation. A meta-analysis described numerous brain regions changed by meditation, which are, for example, 'meta-perception (frontopolar cortex/BA 10), exteroceptive and interoceptive body awareness (sensory cortex and insula), memory consolidation and reconsolidation (hippocampus), self- and emotion regulation (anterior and middle cingulate cortex; orbitofrontal cortex), intra- and interhemispheric communication (superior longitudinal fasciculus; corpus callosum)'.[7]

Thinking and cognitive activities not only depend on spiritual activity, that is, on 'inner light', but on outer light as well. There are studies showing that sunlight not only tans our skin and stabilizes our bones, but also has an effect on our cognitive abilities: lack of sunlight causes a decrease in cognitive powers.[8] However, it is not only external light that has this effect, but also the inner light of meditative work. Thinking has many qualities of light. We speak of 'shining a light on' something, that 'a light goes on', or, being 'enlightened'. Meditative thinking has an inner light quality. It has a healthy effect on the nervous system, right into our neuroanatomy.

Meditation and the Circulatory / Cardiovascular System

The health effects of meditation on the cardiovascular system have long been known. Dean Ornish and others showed almost 30 years ago how meditation and other lifestyle changes have a healthy influence on the major chronic diseases of the cardiovascular system, such as coronary heart disease. In addition, we can develop healing powers through active, spiritual development. We also know the great importance of *feelings*, especially for diseases related to heart and blood pressure. To be able to mobilize healing powers, we need a feeling of being connected to other people. Isolation and social distancing lead to loneliness, to feeling alone, which has a negative effect. Depression as the 'lightlessness of the soul' is also a particular risk factor for numerous cardiovascular diseases and diseases leading to sclerosis (coronary heart disease, arteriosclerosis, chronic lung diseases, as well as osteoporosis). The often encountered mental torpor is followed by physical torpor.

Meditation and the Metabolic System

The third area concerns the metabolic system, its regenerative functions and other biochemical and immunological processes. How are these otherwise unconscious processes connected to the spiritual and soul being of the human being? This very intriguing question has been posed again and again for over 400 years. The connection between consciousness and life has remained something of a mystery since Descartes' dualistic model. Meditation has repercussions on certain metabolic processes, such as chronic inflammation, which are associated with many diseases of our time. Depression, diabetes, cardiovascular disease, but also cancer, are accompanied by chronic inflammation. They form the topsoil on which these diseases develop.

Trauma that a person has experienced in childhood and adolescence is imprinted on their physiology and may manifest visibly as chronic inflammation in adulthood. Medicine therefore needs a biographical dimension for understanding the illness and then the therapy. Meditation can positively influence the course of chronic inflammation. These examples thus point to the connection between consciousness and the life processes of the metabolic system.

Cell Division and Telomerase—What Influences Them

There are interesting observations on this from the last century. Researchers devoted themselves to the question of what has a positive effect on replication and cell division, on the proliferation of life. It was known that with each division our genetic material is 'shortened', so to speak; the ends of the chromosomes, the telomeres, are shortened—a process that is also related to ageing and senescence. The first research on this goes back to the later Nobel Prize winner, Hermann Müller, who described this connection for the first time in 1938. He speaks of an enzymatic event that counteracts the negative effect of cell division and is thus linked to the regeneration of life processes. This enzyme activity was later described as telomerase activity.

Effects on the Healing Powers of the Human Being

Elisabeth Blackburn researched this topic together with others, for which her research group also received the Nobel Prize in 2009. She studied ciliates, which have high telomerase activity. In 2000, Blackburn met a psychologist. They studied mothers of chronically ill children who lived in constant worry about their child and suffered severe psychological stress. The longer the mothers cared for their sick children and the more they suffered, the shorter their telomeres were. Permanent tension and stress significantly reduces the life processes of regeneration, that is, telomerase activity. Life strains, stress and biographical challenges are inscribed at the epigenetic level of the living.

Connection Between Consciousness and Life

This raises the question of the connection between consciousness—in the sense of excessive stress in this context—and reduced regenerative life processes. The connection between life and consciousness has been known for a long time. It was formulated by an important philosopher Karl Fortlage (1806-1881) as early as 1860. Our consciousness results in degradative forces; our sleep, so when we are not conscious, is associated with restorative forces. Fortlage uses a beautiful image that every bit of consciousness in the organism represents a partial death, the death of the human being is accordingly accompanied by increasing consciousness. This opposing connection between consciousness on the one hand and diminution of the living on the other was elaborated by Rudolf Steiner. He described the forces of thought as metamorphosed life forces.[9] A deeper understanding of the living thus emerges: on the one hand, life can unfold in the organism and lead to regenerative formative living phenomena. Conversely, the same forces are found on another level, namely in the world of consciousness of the human being, and can become creatively and formatively effective. Thus the connection between consciousness and life, which is decisive for Waldorf education, medicine and the support of people in general, spans more than 150 years and is given new relevance by modern research results.

Connection Between Meditation and Life

Now, everyday consciousness, with its vitality-draining activity, is the opposite of meditative activity. Stressed consciousness, when there is too much information input, goes hand in hand with degradative processes. Creative spiritual activity, which develops life forces in thinking and not losing them in the abstract world of thought, strengthens regenerative life processes and therefore has a healing effect. Meditation can therefore not only positively influence the nervous system and the heart-lung system, but also supports the regenerative life processes and, for example, increases telomerase activity. In the interim, further studies have proven the connection between meditation (but also lifestyle changes and nutrition) and telomerase activity.[10,11,12] Meditative life promotes life processes, while stressful information-intake through our everyday consciousness weakens the body.

The special feature of meditative consciousness is that spiritual content and qualities connect to the human being. Life is full of wisdom and we marvel at its complex, well-ordered functionality and its mutually co-ordinating relationships. Through thinking, we recognize these natural laws and functional relationships. They are connected to it, indeed they are of the same nature, by initially working undetected in the living organism and then by becoming conscious through thinking. If thinking connects with the spiritual world of truth, this wisdom-filled order of the living is strengthened. Conversely, untruth and lies lead to their impairment. Meditation therefore not only means 'coming to rest', but lives from the connection to actual spiritual content, from the relationship to this world of truth, to which the marvellous beauty of living things and the functional selflessness of its organs originate; 'for the "I" receives its nature and significance from that with which it is bound up,'[13] as Rudolf Steiner expressed it. In this respect, *word* meditations, through working inwardly with verses (mantras), are of great importance in anthroposophical meditation. Rudolf Steiner applied the healing effects of meditation in practice as early as 1920 and recommended its therapeutic use, known as the meditations for patients.[14]

Meditation can therefore make a profound contribution to increasing healing and regenerative powers. It depends on this special quality of thinking, which does not die in the 'cold light' of abstract information, but develops its life and its warmth in this spiritual activity of selfless devotion. In this process, thinking is freed from abstraction, acquires warmth of heart, transforms head thinking to heart thinking. However, meditation does not replace the need for therapy. Rather, in the overall concept of anthroposophic-integrative medicine, it is a contribution that merges with other treatments and medicinal therapy.

What strengthens the soul?

The soul can inhabit the body in such a way that disease processes develop. Stress, agitation, anxiety and depression have negative bodily consequences in this sense. If, on the other hand, the soul comes 'back to its centre', it can promote healing and health. Therefore, which qualities of soul weaken the disposition and make a person susceptible to viral infections such as the Corona virus? In particular, five aspects—which have a pathogenetic influence, but conversely, also indicate in which direction the healing forces can be sought—are essential here.

Cultivate Attentiveness and Concentration

Too much information has a weakening effect on the immune system. A lot of unprocessed information creates stress. Having too many Facebook contacts, for example, can have an immunosuppressive effect. There is now talk of 'Facebook-induced stress' which has shown to have an influence on respiratory infections.[15] Thinking in itself has a light-bestowing benefit for the immune system, but in the case of information overload, it loses its life and dissipates, thus having the opposite effect. Multitasking seems to go in a similar direction.[16]

Develop Self-effectiveness and Autonomous Action

When we create something out of our own initiative, when we take responsibility for putting a decision into action, we experience an increase in strength, a boost. On the other hand, negative health effects arise from externally driven demands on us, and from working under pressure. In pyramidal-hierarchical companies, where the upper echelons give instructions to the lower ones and create pressure, the health of the employees suffers—as a study from Sweden on cardiovascular risk factors has proven.[17] We therefore need to cultivate a culture of 'will', of self-motivated, self-directed and self-responsible action.

Deal with Fear and Stress

Fear, anxiety, agitation and stress have a negative effect on the immune system. The immunosuppressive effects of stress and mental strain (something which patients often complain about) have been known for a long time[18,19,20] and contribute to a susceptibility to infections. While the one-sided dominance of the consciousness pole (nerve-sense system) is associated with susceptibility to infections, moderate (non-exhaustive) exercise, thus activity of the metabolic-limb system, seems to have a positive influence on the immune system.[21]

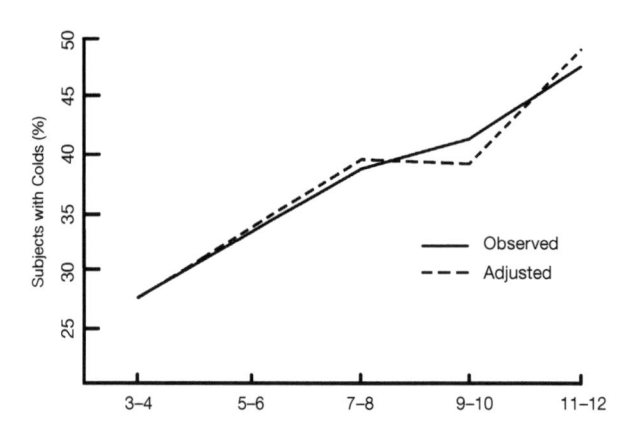

Relationship between mental stress ('stress index') and susceptibility to infection (percentage of symptomatic test subjects with viral infection of the upper respiratory tract (modified[22]).

This makes it necessary to develop a feeling that between you and the other, the world, you are able to breathe, and this will help you build a bridge to the other. Fear- and anxiety-inducing media messaging, in this light, has a negative effect and worsens immunological function.

Positivity and Interpersonal Relationships

Are we able to find the being of the other person and build a bridge to him or her? This is about the question of interpersonal relationship and its healing powers for the soul. Many people are lonely at the moment—because of lockdowns, but also because of the fear of contagion. Experiencing loneliness is a pathogenic, disease-causing factor.[23] Escaping it and its effects is only possible when our gaze turns to the beautiful and the good in the world. We know from a recent study that the perception of beauty, for example in art—even if only receptively as a viewer, listener or spectator—has a life-prolonging effect on people.[24] The encounter with beauty develops healing powers, a connection that Rudolf Steiner pointed out more than 100 years ago:

> Whenever man regards a thing as being true, beautiful and good, not through dispassionate, intellectual reflection but by a direct encounter, a quickened pulse makes him conscious of the heart's assent. The heart actually beats differently in response to the beautiful than in response to the ugly or pernicious. In this original logic of the heart there is something that may be called spontaneous sympathy. When this logic of the heart which functions in the subconscious becomes more clearly articulate, the heart shows quite plainly by the circulation of the blood that it is an expression of this logic. And a painful experience repeatedly brought before our eyes can influence our bodily nature by way of the heart to the point of causing actual illness. There can be physiological confirmation of this.[25]

The positive health benefits of positivity, a journey of discovery to real and existing positivity not shrouded or obscured by the shadow of negatives of events of the times, are well documented. A number of large epidemiological studies show that a positive outlook on life has a protective effect on chronic non-communicable diseases. Positive psychological well-being seems to be crucial, not only for various cardiovascular problems, but also for resilience in a broader sense. Making life meaningful and positive reduces risk factors for chronic heart disease and hypertension. Therefore, a positive attitude towards life has significant benefits for survival in both healthy and sick people.[26]

Unbiasedness Towards Life—Biographical Perspectives
We live from hope and perspective. They are essential for survival; our times clearly illustrate this. Many COVID patients have lost their livelihoods and no longer have any economic prospects. Our society is divided between people who are threatened existentially to the point of starvation, while others notice little of it and hardly understand what is happening to their fellow human beings. And then there are those who are profiting from the current situation.

The theme of 'hope' plays a significant role in medicine. In 2020, numerous scientific papers were published on this topic. Activating the forces of hope as healing forces is currently an important theme. We may not deprive anyone of hope—this is a fundamental principle of the therapist towards the patient. Vaclav Havel expressed it like this, 'Hope is not the conviction that something will turn out well, but the certainty that something makes sense, no matter how it turns out.' He thus echoes people like Victor Frankl, who asked the question of meaning very emphatically out of his own biographical situation. Frankl, founder of logotherapy, said that we need not only analytical psychotherapy and psychology, not only behavioural therapy, but especially a 'height psychology',[27] something that brings light from the heights, that shines into our biography. In order to be able to perceive this light at all, we have to open ourselves to what is to come, having trust in the forces that may manifest as life-giving meaning.

Development Paths of the Soul

With these five aspects of the being human, a five-pointed star emerges that is of great importance for inner development. Each of its five rays represents developmental tasks for the human being:

1. *How can I strengthen concentration in my thinking, which not only gathers information, but actively develops the ability to concentrate?*

We need this strength not only for a hygienic handling of the diverse amount of information coming at us, asked for and unasked for, but in meditative practice as well.

2. *How do I develop from a recipient of orders to my will to a creator of actions, someone who transforms a decision into deeds arising out of inner insight, and in so doing, develops decisiveness and willpower?*

3. *How do I centre myself in my feelings?*

How can my soul's equilibrium be found in the diverse forces around me and how can my own feelings be developed into a kind of sensory organ, not only for feeling myself but for learning to empathize with the other, and feel what is going on in their world?

Especially in times of isolation, we need to develop our feeling life. Rudolf Steiner used a very appropriate term in this regard: healthy intuition.[28] Through feeling, something is intuited, which can then be verified by other methods. Many great inventors had a hunch, or, intuition, before they figured out the exact details of their invention. The opening of our feeling life to the spiritual is something very decisive. It leads the way out of tension, anxiety and inner turmoil. Wonderful images, such as a starry sky, or a sunrise, a sunset, the tranquillity of the sky, all these bring calming forces into the soul, as meditative images for the cultivation and development of our feeling life.

4. *What qualities are required for maintaining or developing positivity in light of our current time, especially in relation to our fellow human beings?*

For building a bridge to the other person, we need to practise 'seeking the positive'. Someone who may be perceived as impossible has a positive core that can quickly be shrouded in antipathy and emotional rejection. Here, an exercise would be to discover this good core and search for it until it is found. In this way, human relationships deepen, affection, compassion and even devotion develops.

The fifth exercise relates to the question:

5. *How can unbiasedness and openness to life be developed?*

Through not having preconceived thoughts, but by being real seekers, by developing the utmost openness to the world in order to learn new things. The many preconceptions we have warp an unbiased outlook, such as: 'A potentized medicine cannot work because it contains no active substance.' Model concepts restrict an independent perspective and do not allow for other possibilities. The human being is then trapped by 'what cannot be, must not be' (Christian Morgenstern). This exercise is about giving attention to the many thoughts and ideas that limit an unbiased perception.

Paths of Schooling for Inner Development
The Six Attributes

In order to cultivate the aforementioned five aspects of being human, one can begin with the five-pointed star: exercise concentration, develop will and feeling, train positivity, and train an unbiased and an unending willingness to learn new things. In this way, five capacities of soul are created, which arise like five rays from the inner activity in the soul.

They can now be merged into a sixth quality, which forms a circle around the star (see illustration). From the development of thinking, feeling and willing, as the three germinal leaves of the soul, a supersensible organ is formed, which consists of these abilities and learns to perceive the world and other human beings more deeply. In this supersensible organ building, which is achieved by bringing together and combining the five exercises described, there are now six qualities that are decisive for inner development. We acquire them as an I-being which becomes effective in thinking, becomes a leader in will, masters feelings, learns to see the positive in others and learns new things from every moment that life brings.

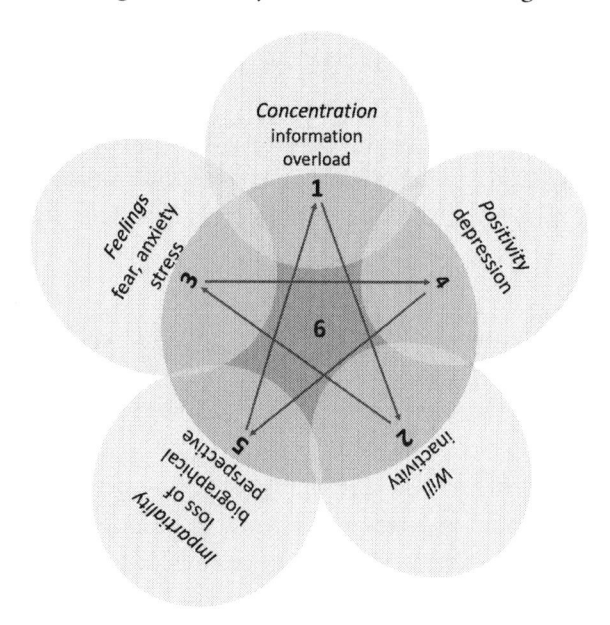

The six attributes:
Exercises to promote the powers of healing through inner development.
Immunocompromising and disease-causing mental stresses

Christian Morgenstern, whose 150th birthday we celebrated in May 2021, summarized these six attributes, as formulated by Rudolf Steiner,[29] in a profound poem.

It is a kind of guideline for the development and strengthening of the soul:

Geschöpf nicht mehr, Gebieter der Gedanken,
Des Willens Herr, nicht mehr in Willens Frohne,
Der flutenden Empfindung Maß und Meister,
zu tief, um an Verneinung zu erkranken,
zu frei, als dass Verstocktheit in ihm wohne:
So bindet sich ein Mensch ans Reich der Geister:
So findet er den Pfad zum Thron der Throne.[30]

Creature no more, master of thoughts,
Lord of the Will, no more will's slave,
The measurer and master of surging sensations,
... too deep to be diseased by denial ..,
... too free for the blockages within him ...:
Thus binds the human to the spheres of the spirit:
And thus finds the path to the throne of thrones.[30]

Qualities have been addressed here, which develop the soul. When this organism of faculties unfolds like a five-pointed star with its unifying circle, it has a healthy influence on the soul and thereby on the body as well.

The Power of the Sun in Us

The shadow reflections of these qualities—overloaded thinking, stress-generating willing, fear and anxiety in feeling, isolation and loneliness, loss of perspective—have limiting and burdening effects on the immune system, whereas, developing these six qualities strengthens our healing forces. In them, the individuality, the 'I' of the human being works by learning to master and direct these capacities of soul.

If you are lucky enough to meet very old people in hospital or in an outpatient department and experience how their personality, their 'I' shines from their eyes, how it appears in their gestures and facial expressions, you get the impression of a sun-like radiance streaming from them. In palliative care too, therapists, nurses and doctors experience this radiance.

Like a sun, their beings manifest light, human warmth, and time and again pure kindness. By schooling these six qualities, these sun forces will strengthen people, leading to 'I' competence and effectiveness, while having a healthy effect in the body at the same time.

Everything we experience in this practice, in terms of feelings of failure, imperfection and defeat, provides us with an inner image of ourselves, a kind of tool for self-diagnosis, which makes clear to us a starting point in terms of practice for further modest steps in our development, so that we become more and more human, a true human being, one step after the other. In this path of practice, the schooling of these six qualities, lives an invitation, a calling, known to us since the ancient mysteries and just as apt today for the development of the human being: *Know thyself!* The path of inner development is connected to health. The path to being more human is connected to our forces of healing.

Connecting to Spiritual Content through Meditation

One particular meditation speaks to these sun forces. There are many different types of meditations: meditations that connect us better to our body, meditations that make us more mindful of our breath. There are also meditations that connect us to spiritual content through inner attention. This meditative activity does not then direct inner attention to the body and its functions, or to soul content, but connects us to the spiritual world to which the essence of the human being belongs and which is our home. This gives rise to a special strength that not only provides rest and relaxation, but is a substance, which has a positive, strengthening influence on us. One of these meditations was given by Rudolf Steiner in 1923:

In meinem Herzen
Strahlt die Kraft der Sonne
In meiner Seele
Wirkt die Wärme der Welt.
Ich will atmen
Die Kraft der Sonne.
Ich will fühlen
Die Wärme der Welt.
Sonnenkraft erfüllt mich
Wärme der Welt durchdringt mich.

In my heart
The might of the sun radiates
In my soul
The warmth of the world works.
I want to breathe
The might of the sun.
I want to feel
The warmth of the world.
Might of the sun fill me,
Warmth of the world permeate me.[31]

The relationship of heart and sun, known from ancient cultures, is now being addressed. The Egyptian Book of the Dead contains the following depiction of the heart: 'You are the pounding in my heart'—this expresses their perception of the sun. Its power, which surrounds us as light, radiates through its expanse of light into the human heart, which we feel in relationship to our 'I' and in which the bodily nature of the human being is centred. The warmth of the world works into our soul, into the feelings connected with our heart and the middle, rhythmic human being. Meditating in this way is always challenging. One does not learn it once and for ever, like riding a bicycle. While in meditative work, making the effort gives life, this spiritual activity of the human being also has a lot to do with grace. Good, and even redemptive thoughts, do not just arise on command. It takes individual effort and then grace-filled moments when the longed-for thought emerges. The same applies to meditation.

Fruits of Meditative Work and Inner Development

What are the fruits of such efforts, which are practised with all our imperfections, for long periods, and accompanied by many failures? Especially the fifth exercise on learning to be open and being prepared for what is approaching, to what might come, is decisive here. For after meditating, it may be that completely new thoughts and sensations arise in the soul. There are everyday things that correspond to this. To illustrate: when you wake up in the morning, feeling tormented by the alarm clock and having to leave your cosy warm bed, and say: 'The early bird catches the worm' may feel a little bewildering, or even alien. A short time later, perhaps in the shower, a plethora of new thoughts may suddenly arise in you.

Studies have looked into the question: At what time of day do new thoughts occur—thoughts that, for instance, centuries later, we still draw on, which have had a significant influence on our culture? For example, the 'inspiration' of Kekulé's ring structure for benzene happened in the morning shortly after waking up, a memory from a dream.[32] The concept of the periodic table is also said to have fallen into place by Dimitri Mendeleyev in the morning hours after a dream.

The human being comes out of the world of sleep, out of the spiritual world, into the life of the new day, and like an echo, good thoughts arise, for musical compositions as well, as we know from musicians like Mozart. This requires self-activity, for no great discovery has simply come out of sleep without prior effort. Through preparation, however, new thoughts can appear in these moments and inspire us. Meditative work has something similar to such a morning mood. When, after its completion, meditation fades away and everyday work begins, moments of grace are noticeable, which give rise to new ideas and good thoughts.

There is an essential prerequisite for meditative life which is—following an indication by Rudolf Steiner—an increased feeling of devotion. Through this force—which has to do with warm interest in other people, in spiritual content, and is connected to the forces of love—a bridge to others is built. An impression can be gained—and this is also very decisive for the ethical-spiritual development of nurses, doctors and therapists in Anthroposophic Medicine—that the fruits of this warm interest is the unfolding of a deepened relationship with the ill person. And finally, a more powerful consequence results when we connect with an idea, in full devotion. When 'a light comes on' inwardly, evidenced by the fact that we feel it, these strong forces become available to us more than when things are merely carried out *comme il faut*.

In this respect, meditation has an affect on all the forces of the soul: on thinking, feeling and will. And—as already mentioned at the beginning—it has an effect on our body in the sense of strengthening our health-giving forces. This rounds off the initial question about the strengthening influence of inner development and meditation on our disposition and its healing capabilities. For dealing with the COVID pandemic, it follows that a responsible approach is needed to reduce exposure, especially in protection of risk groups, a differentiated treatment of the COVID disease in its four stages[33], and the inclusion of the soul and spiritual spheres for strengthening resilience. The promotion of salutogenetic forces needs an appropriate lifestyle, care of personal relationships in times of 'social distancing', as well as inner development.

Written manuscript of a lecture given at the Goetheanum on 21 December 2020.

1. Von Goethe, Johann Wolfgang; Eckermann, Johann Peter. *Conversations of Goethe with Johann Peter Eckermann* (pp. 463-464). Kindle Edition.
2. Cohen, S., Tyrrell, D. A., Smith, A. P.: 'Psychological stress and susceptibility to the common cold.' *New England Journal of Medicine* 1991; 325 (9). S. 606–612.
3. Pedersen, A., Zachariae, R., Bovbjergm, D. H.: 'Influence of psychological stress on upper respiratory infection—a meta-analysis of prospective studies.' *Psychosomatic Medicine* 2010; 72 (8). S. 823–832. https://insights.ovid.com/pubmed?pmid=20716708 (Accessed May 2019.)
4. Kalula, S..; Ross, K.: 'Immunosenescence-inevitable or preventable?' *Current Allergy & Clinical Immunology* 2008; 21. S. 126–130.
5. Black, D. S, Slavich, G. M.: 'Mindfulness meditation and the immune system: a systematic review of randomized controlled trials.' *Ann N Y Acad Sci.* 2016 Jun; 1373(1):13–24.
6. Seifert, G., Jeitler, M., Stange, R., Michalsen, A., Cramer, H., Brinkhaus, B., Esch, T., Kerckhoff, A., Paul, A., Teut, M., Ghadjar, P., Langhorst, J., Häupl, T., Murthy, V., Kessler, C. S.: 'The Relevance of Complementary and Integrative Medicine in the COVID-19 Pandemic: A Qualitative Review of the Literature.' *Front Med* (Lausanne). 11 Dec. 2020; 7:587749.
7. Fox, K. C., Nijeboer, S., Dixon, M. L., Floman, J. L., Ellamil, M., Rumak, S. P., Sedlmeier, P., Christoff, K.: 'Is meditation associated with altered brain structure? A systematic review and meta-analysis of morphometric neuroimaging in meditation practitioners.' *Neurosci Biobehav* Rev. 2014 Jun; 43:48–73.
8. Kent, S.T., Kabagambe, E. K., Wadley, V. G., Howard, V. J., Crosson, W. L., Al-Hamdan, M. Z., Judd, S. E., Peace, F., McClure, L. A.: 'The relationship between long-term sunlight radiation and cognitive decline in the REGARDS cohort study.' *Int J Biometeorol.* 2014 Apr; 58(3):361–70.
9. Steiner, R., Wegman, I.: *Fundamentals of Therapy. An Extension of the Art of Healing through Spiritual Knowledge.* Chapter 2, 'Why Does Man Become Ill?' (GA 27).
10. Ornish, D., Lin, J., Daubenmier, J. et al.: 'Increased telomerase activity and comprehensive lifestyle changes: a pilot study.' *Lancet Oncology* 2008; 9 (11). S. 1048–1057.
11. Jacobs, T. L., Epel, E. S., Lin, J. et al.: 'Intensive meditation training, immune cell telomerase activity, and psychological mediators.' *Psychoneuroendocrinology* 2011; 36 (5). S. 664–681.
12. Lavretsky, H., Epel, E. S., Siddarth, P. et al.: 'A pilot study of yogic meditation for family dementia caregivers with depressive symptoms: effects on mental health, cognition, and telomerase activity.' *International Journal of Geriatric Psychiatry* 2013; 28 (1). S. 57–65.
13. Steiner, R.: *Theosophy*, 'Chapter 1, The Nature of Man, 4. Body, Soul and Spirit.' (GA 9)

[14] Selg, P.: *Patienten-Meditationen von Rudolf Steiner.* Verlag des Ita Wegman Instituts, Arlesheim 2019. [Patient Meditations given by Rudolf Steiner.] Not yet translated into English.

[15] Campisi, J., Bynog, P., McGehee, H., Oakland, J. C., Quirk, S., Taga, C., Taylor, M.: 'Facebook, stress, and incidence of upper respiratory infection in undergraduate college students.' *Cyberpsychol Behav Soc Netw.* 2012 Dec; 15(12):675–681.

[16] Modi, H. N., Singh, H., Darzi, A., Leff, D. R.: 'Multitasking and Time Pressure in the Operating Room: Impact on Surgeons' Brain Function.' *Ann Surg.* 2020 Oct; 272(4):648–657.

[17] Wamala, S. P., Mittelmann, M. A., Horsten, M. et al.: 'Job stress and the occupational gradient in coronary heart disease risk in women. The Stockholm Female Coronary Risk study.' *Social Science and Medicine* 2000; 51. S. 481–489.

[18] Cohen, S., Tyrrell, D. A., Smith, A. P.: 'Psychological stress and susceptibility to the common cold.' *New England Journal of Medicine* 1991; 325 (9). S. 606–612.

[19] Pedersen, A., Zachariae, R., Bovbjergm D. H.: 'Influence of psychological stress on upper respiratory infection—a meta-analysis of prospective studies.' *Psychosomatic Medicine* 2010; 72 (8). S. 823–832.

[20] Kalula, S.; Ross, K.: 'Immunosenescence-inevitable or preventable?' *Current Allergy & Clinical Immunology* 2008; 21. S. 126–130.

[21] Brolinson, P. G., Elliott, D.: 'Exercise and the immune system.' *Clinics in Sports Medicine* 2007; 26 (3). S.311–319. (Accessed May 2019.)
https://www.sportsmed.theclinics.com/article/S0278-5919(07)00025-7/fulltext

[22] See endnote 18.

[23] Valtorta, N. K., Kanaan, M., Gilbody, S., Hanratty. B.: 'Loneliness, social isolation and risk of cardiovascular disease in the English Longitudinal Study of Ageing.' *Eur J Prev Cardiol.* 2018 Sep; 25(13):1387–1396.

[24] Fancourt, D., Steptoe, A.: 'The art of life and death: 14 year follow-up analyses of associations between arts engagement and mortality in the English Longitudinal Study of Ageing.' *BMJ.* 2019 Dec 18; 367:l6377.

[25] Steiner, R.: *Macrocosm and Microcosm.* 'Transformation of Soul-Forces and Stages in the Evolution of Physical Organs, Reading of the Akashic Chronicles.' Lecture on 30 March 1910. (GA 119)

[26] See endnote 11.

[27] Frankl, V. E.: Wer ein Warum zu leben hat. Lebenssinn und Resilienz. Beltz Verlag, Weinheim 2017, S. 37. [Translator's note: There are many translations into English of Frankl's work, but I cannot find the particular one referred to here. However, 'height psychology' is mentioned in several of his works.]

[28] Steiner, R.: *An Outline of Occult Science.* Chapter V, 'Cognition of the Higher Worlds. Initiation.' (GA 13)

[29] Ibid.

30 Morgenstern, C.: Werke und Briefe. Bd. II: Lyrik 1906–1914. Verlag Urachhaus, Stuttgart 1992, S. 240. Translation of the poem by C. Howard.
31 Steiner, R: Mantric Sayings Meditations. 1903–1925. (CW 269). Translation of the German verse by C. Howard.
32 The ring structure of carbon atoms in organic molecules: According to legend, this idea came to Kekulé in a dream, in which, snakes held onto each other's tails, creating a ring structure.
33 Soldner, G., Breitkreuz, T.: COVID-19.
https://www.anthromedics.org/PRA-0939-EN#list-sections-4

What are the Intentions of the School of Spiritual Science at the Goetheanum?

The Goetheanum in Dornach is the centre of the School of Spiritual Science. This was founded by DPhil. Rudolf Steiner (1861–1925) and his co-workers with the aim of fostering training and study, further education and practical initiatives in various aspects of civilization. The founding of this school was based on the insight that an exclusively reductionistic scientific approach cannot effectively address the complex problems of the individual areas of life. Rather, it was considered necessary to bring about holistic, ecologically rational concepts for a productive connection between the natural sciences and humanities, art and religion. From the very beginning, a new understanding of the living and developmental processes in nature and in human beings was essential, one which must be understood, protected and promoted. Rudolf Steiner developed an in-depth research approach that opens up important perspectives in this regard, through anthroposophy as a modern science of the spirit.

The School of Spiritual Science at the Goetheanum, which is separated into different specialized sections, has included a path of meditative self-development since its foundation. This is not considered an end in itself, but as an integrated, necessary and indispensable part of a responsible attitude towards research, life and work. It is about methodically unfolding one's own soul forces, that is, through strengthening the 'I' of the human being and learning to consciously manage it in its differentiated relationships to the natural and spiritual world.

The path of meditative self-development and the concrete inner training of the human being is of fundamental importance in all areas of the work of the Goetheanum. Art also plays a special role in the overall context of the school. The Goetheanum has productive working spaces in which the innovative power of anthroposophy can be experienced and viewed through various artistic disciplines. In addition, artistic training components play an essential role in all the sections of the School of Spiritual Science, since they foster presence of mind and the ability to act intuitively and proactively, enabling creative handling of problems and social cooperation.

The scientific and artistic research activities, as well as the training and further education options at the Goetheanum, are organized in a decentralized manner. In contrast to many other schools of higher education, the work is not only carried out on site, but also through international courses, teaching and further educational events, with the participation of the Dornach collegium of the School of Spiritual Science. For instance, further training courses in Anthroposophic Medicine and curative education, in Waldorf education and biodynamic agriculture, as well as training courses and projects in the artistic spheres of work, are held all over the world and are co-organized by the individual sections at the Goetheanum. Research initiatives and projects in all these areas are also coordinated and supervised worldwide.

In regular conferences and meetings of the relevant professional groups at the Goetheanum, with participants from all continents and simultaneous translation into up to seven languages, results are presented, experiences are jointly reflected upon, systematized and presented to the public through corresponding publications and presentations. In this way, results gained from practical experience find their way into the ongoing training and research work of the School of Spiritual Science. In terms of content, this work is thus also supported and shaped to a considerable extent by international cooperation. At the Goetheanum itself there are also introductory and advanced courses in anthroposophical spiritual science. The combination of research, teaching and practice, of individual and community responsibility and of inner training and cosmopolitan social orientation, distinguishes the Goetheanum in Switzerland as a place of higher learning.

In Anthroposophic Medicine, education, agriculture, anthroposophical art and other fields, the Goetheanum's research approaches and results have proven to be constructive, life-enhancing, ecological and sustainable. This has led to contacts and constructive working contexts, also with numerous non-anthroposophically-oriented professional groups, representatives of civil society, scientists and university lecturers. The need for cooperative projects for the protection and preservation of endangered life in both ecological and humanistic aspects is becoming increasingly clear to society. There is also growing realization that these goals can only be achieved through an extraordinary scientific effort and cooperation, not only across disciplines but also across different paradigms.

The work and finances of the School of Spiritual Science are supported by the members of the Anthroposophical Society all over the world, but its international projects are also supported by private donors and charitable foundations.

About the Authors

Paula Edelstein, born 1971, Argentina. Anthropologist. Waldorf teacher and lecturer at secondary and tertiary level at the Escuela e Instituto de Formación Docente Perito Moreno, the Asociación Civil Educadora Luz del Sol. Member of the research programme 'Relationships between specific didactics and cultural identities: an interdisciplinary approach to their analysis', in collaboration with a team from the Didactics of Science and another from the Didactics of Geography (Universidad Nacional de Luján-UNLu. Department of Education, in cooperation with the Asociación Luz del Sol).

Jean-Michel Florin studied agriculture and nature conservation in France and Goethean Science at the Research Institute at the Goetheanum. He gives many courses and is active in various networks on environmental and social issues. He is currently coordinator of the French Biodynamic Association, *Mouvement de l'Agriculture Bio-Dynamique*, in Colmar and co-leader of the *Section for Agriculture* at the Goetheanum. He has produced various publications on biodynamic themes.

Dr med. Matthias Girke is co-founder of the Gemeinschafts-krankenhaus Havelhöhe, a clinic for anthroposophical medicine in Berlin, where he was head physician of General Internal Medicine for over 21 years. Today he is still active in an advisory capacity and in the clinic's outpatient department. In September 2016, Matthias Girke took over the Leadership of the *Medical Section* of the School of Spiritual Science at the Goetheanum, and since April 2017, has been a member of the Executive Council of the *General Anthroposophical Society*.

Gerald Häfner, born in Munich, was co-founder of the Green Party, founder of the organizations Mehr Demokratie e.V., Democracy International, as well as the Petra Kelly Foundation. He was a member of the German Bundestag and the European Parliament for many years. Since October 2015 he has been the Leader of the *Section for Social Sciences* at the Goetheanum.

Dr phil. Christiane Haid is Leader of the *Section for Literary Arts and Humanities* and since 2020, of the *Visual Arts Section* at the School of

Spiritual Science, Goetheanum. Her doctoral thesis was entitled. 'The Myth of Dream and Imagination—Albert Steffen's Little Myths' (Basel 2012) and she has several publications on anthroposophical, historical and literary themes.

Prof. Stefan Hasler grew up on Lake Constance and in Dornach; studied music in Basel and Budapest; studied conducting in London; studied eurythmy in The Hague and Hamburg; taught at the Waldorf School Hamburg-Wandsbek, lectured at the Eurythmy School Hamburg; since 2003, professor of eurythmy at Alanus University; since 2014, Leader of the *Section for Performing Arts*; since 2018, member of the Goetheanum Eurythmy Ensemble; research work in tone eurythmy, sound eurythmy, Raphael and educational of eurythmy.

Ueli Hurter is Co-Leader of the *Section for Agriculture* at the Goetheanum and was co-opted into the Executive Council of the *General Anthroposophical Society* in 2020. Since 2019, member of the Board of Directors of Weleda AG. At the same time, is a farmer, and was in charge of L'Aubier, a Demeter farm, eco-hotel, organic restaurant and an eco-living quarter. From 1997–2010, he was president of the Swiss Demeter Association and co-initiator of 'Sowing the Future!'

Dr Constanza Kaliks, born in Chile and raised in Brazil. Studied mathematics and Waldorf education, doctorate in education with her dissertation on Nicolaus Cusanus. Worked for 19 years as a Waldorf teacher and lecturer in teacher training in São Paulo. Since 2012, Leader of the *Youth Section* at the Goetheanum and since Easter 2020, Co-Leader of the *General Anthroposophical Section*. Since 2015, member of the Executive Council of the *General Anthroposophical Society*.

Florian Osswald, born in Basel, Switzerland, studied process engineering. After training as a curative educator at Camphill, Scotland, attended the teacher-training seminar in Dornach. For 24 years, taught mathematics and physics at the Rudolf Steiner School in Bern and Ittigen and worked as a collegial advisor for various countries. Since the beginning of 2011, is Co-Leader of the *Pedagogical Section* at the Goetheanum with Claus Peter Röh.

About the Authors

Dr rer. nat. Matthias Rang is a physicist. Since 2007, at the Research Institute and since 2020, is Co-Leader of the *Natural Science Section* at the Goetheanum.

Claus-Peter Röh studied education, and from 1983, worked as a class, music and religion teacher at the Freie Waldorfschule in Flensburg. Since 1998, member of the Initiative Circle of the *Pedagogical Section in Germany*. Since January 2011, Co-Leader of the *Pedagogical Section* at the Goetheanum in Dornach with Florian Osswald and since Easter 2020, Section Leader of the *General Anthroposophical Section* at the Goetheanum in Dornach.

Prof. Dr med. Peter Selg is a specialist in child and adolescent psychiatry and psychotherapy. He teaches medical anthropology and ethics at the Alanus School of Spiritual Science in Alfter and at the University of Witten/Herdecke. He directs the Ita Wegman Institute for Basic Anthroposophical Research in Arlesheim and since Easter 2020, works with the leadership of the *General Anthroposophical Section* at the Goetheanum.

Georg Soldner, after eight years of clinical training at Munich's paediatric hospital (*Münchner Kinderklinike*), became a resident specialist in paediatric and adolescent medicine. He is currently Deputy Leader of the *Medical Section* of the School of Spiritual Science at the Goetheanum. He is also the leader of the *Academy of Anthroposophical Medicine of the Society of Anthroposophic Physicians* in Germany and is publisher and editor of the *Vademecum* of Anthroposophic Medicines.

Dr Johannes Wirz, Ph.D., is a molecular biologist, Co-Leader of the *Research Institute* and since 2020, Co-Leader of the *Natural Science Section* at the Goetheanum.

Justus Wittich, born in Berlin, studied economics and law at the Freie Universität Berlin. Since 1979, editor of various journals and from 1985 to 2018, managing director of the educational institution *der hof* in Frankfurt-Niederursel. Since 2012, on the Executive Council of the *General Anthroposophical Society* and is its Treasurer.

A NOTE FROM RUDOLF STEINER PRESS

We are an independent publisher and registered charity (non-profit organisation) dedicated to making available the work of Rudolf Steiner in English translation. We care a great deal about the content of our books and have hundreds of titles available – as printed books, ebooks and in audio formats.

As a publisher devoted to anthroposophy...

- We continually commission translations of previously unpublished works by Rudolf Steiner and invest in re-translating, editing and improving our editions.

- We are committed to making anthroposophy available to all by publishing introductory books as well as contemporary research.

- Our new print editions and ebooks are carefully checked and proofread for accuracy, and converted into all formats for all platforms.

- Our translations are officially authorised by Rudolf Steiner's estate in Dornach, Switzerland, to whom we pay royalties on sales, thus assisting their critical work.

So, look out for Rudolf Steiner Press as a mark of quality and support us today by buying our books, or contact us should you wish to sponsor specific titles or to support the charity with a gift or legacy.

office@rudolfsteinerpress.com
Join our e-mailing list at www.rudolfsteinerpress.com

RUDOLF STEINER PRESS